AN INTRODUCTION TO MATERIALS AND CHEMISTRY

W0113650

This new edition of *An Introduction to Materials and Chemistry*, book 1 in the updated *Science for Conservators* series, provides conservators and conservators-in-training with a very basic introduction to the language of chemistry and to the scientific approach.

Drawing on 40 years of experience as a conservation scientist, Joyce H. Townsend takes readers through the elementary steps that will enable them to understand and investigate materials in historic objects, and those modern materials used to conserve them, in scientific terms. The book also introduces basic chemistry concepts. It provides worked examples and exercises throughout. This new edition has been significantly expanded and updated, with new material about health and safety, sustainability, and the trend to use greener materials, amongst other topics. The book also includes all-new illustrations, a list of further reading and is accompanied by a Companion Website, which features additional examples, illustrations and more.

An Introduction to Materials and Chemistry assumes no previous scientific knowledge and will be essential reading for pre-program applicants to, and students already on, postgraduate conservation programs worldwide. It will also be useful to conservators who are looking to refresh their knowledge or to fill gaps in their training, and for those who trained in languages other than English, but now work in that language.

Joyce H. Townsend, FIIC, ACR, is series editor for the updated *Science for Conservators* series. She has been IIC Director of Publications for over ten years, and a conservation scientist for over four decades, more than three of them spent at Tate in London, UK, where she specialises in microscopical methods for the analysis of paint, and the interpretation of the techniques of British artists working in oil and watercolour, as well as microfading and X-ray fluorescence studies for works on paper. She has published widely for a range of audiences. Since 2019 she has been an honorary professor in the School of Culture and Creative Arts, University of Glasgow, UK.

Science for Conservators
Series editor: Joyce H. Townsend

The Science for Conservators series is now being revised and fully updated, and it will be extended in scope. The original series has provided key basic texts for conservators throughout the world since its publication in the late 1980s. Scientific concepts are basic to the conservation of artefacts of every type, and these introductory volumes provide an essential theoretical background for conservators who have entered the field without scientific training. It is designed for pre-programme, student and personal study, and also serves as a checklist of scientific terms for those working in English as a second or third language.

An Introduction to Materials and Chemistry
Book 1
Joyce H. Townsend

For more information about this series, please visit: www.routledge.com/Science-for-Conservators/book-series/SFC

An Introduction to Materials and Chemistry

Book 1

Third Edition

Joyce H. Townsend

Routledge
Taylor & Francis Group

LONDON AND NEW YORK

Designed cover image: © amtitus / Getty Images

Third edition published 2024
by Routledge
4 Park Square, Milton Park, Abingdon, Oxon, OX14 4RN

and by Routledge
605 Third Avenue, New York, NY 10158

Routledge is an imprint of the Taylor & Francis Group, an informa business

First edition published by the Crafts Council 1983
Second edition published by Routledge 1992

British Library Cataloguing-in-Publication Data
A catalogue record for this book is available from the British Library

ISBN: 978-1-032-20010-1 (hbk)
ISBN: 978-1-032-20008-8 (pbk)
ISBN: 978-1-003-26186-5 (ebk)

DOI: 10.4324/9781003261865

Typeset in Times New Roman
by codeMantra

Contents

Figures

Tables

Foreword and acknowledgements

Several anonymous reviewers were asked to comment on my proposal for updating this book, and they were provided with the original text so that they could point out areas in need of updating, or any inaccuracies that had crept into the first edition. This author then repeated the checking process for any inaccuracies, obsolete topics and for explanations that seemed too short to be meaningful. That is why some of the original content does not appear in this edition. The input of the reviewers has assisted both with the selection of new topics and the shaping of the whole expanded series and I am grateful to them, and especially to the one reviewer who volunteered to and later critically reviewed the first draft of the updated text. Some of the reviewers are also teachers of conservation, and several sent examples of the reading lists they provide for their own students: I have used some of their suggestions in the section on further reading and added my own suggestions too. I should also thank my own colleagues, the students in a number of conservation programmes where I have given occasional lectures, and the authors of the hundreds of papers I must have reviewed and edited over the years: their actual questions, written texts, and what they omitted from their texts as much as what they wrote have all indirectly fed into the contents of this book.

This book, the first in the series, assumes no previous scientific knowledge at the start. Science builds up knowledge that leads to understanding, theories, and then predictions from theories that can be tested, one step at a time: if you try to read a later section of the book in advance of others, you will run the risk of becoming very confused, or else of only partially grasping its meaning.

Remember that *Book 1* on materials is far from being a complete scientific course in itself. It will be necessary to read the other two books in the original series, *Cleaning* and *Adhesives and Coatings* (whose updated texts are being planned now) before a useful vocabulary and understanding is built up on the process of looking at an object, formulating a treatment plan or deciding that no direct intervention could be the best option, setting about cleaning it, and setting about dealing with damages and areas of weakness that it has accumulated. You may also find that the order of this book varies slightly from more standard science textbooks, but this is because the text is structured to suit the specific needs of trainee and emerging conservators.

Book 1 provides a very basic introduction to the language of science and to the scientific approach. In fact, it mostly covers chemistry, and the subject taught in some UK universities as physical chemistry. It takes you through some crucial elementary steps towards being able to identify materials in scientific terms and introduces you to basic chemistry, but not to basic physics. It does not discuss conservation treatments, nor the selection of chemicals to use in the conservation studio/laboratory, but it does outline the chemistry behind some classes of artefacts. Gradually, as the series moves on, the science taught in this book will be developed further, as the science behind different conservation procedures is discussed. The final chapter of this first book in the series will also provide you with a useful guide to the chemical names frequently encountered in conservation now and in the recent past, showing how their chemical properties are related to their structures.

Foreword to the *Science for Conservators* series

The series was first published in 1982 as a series of three books aimed at conservation students and conservators who had come into the field without any scientific training beyond their high school years. It was intended to develop into some six or seven topics covering a broader spectrum of knowledge than the first three, which concentrated greatly on chemistry. The series of three was reprinted in 1987 and produced by Routledge as a second edition in 1992. It has been in steady demand ever since, despite the text being unchanged since the early 1980s. Some conservation training programmes in a range of countries have recommended the series for pre-programme study, while others have included it in their first year reading list for the students from a humanities background or for all their students.

Having no training in the natural sciences after the early teenage years, and sometimes even fewer encounters with the biological sciences, had been the background of most conservators in all countries for much of the twentieth century. It still is for some pre-programme conservation students from a few countries, for those who undertake apprentice-style practical training straight after high school and for others who enter the field of cultural heritage conservation through postgraduate training that was preceded by humanities-based or social sciences-based graduate study. The conservation of artefacts is a truly inter-disciplinary activity which would best be begun with the knowledge of a lifetime! Most conservation training programmes today, including those for heritage scientists, aim to provide pathways into thought processes and knowledge bases from the natural sciences, the humanities generally and to a lesser extent the social sciences, engineering, the biological sciences and philosophy and ethics as well, through targeted teaching in all these areas.

The fully revised and updated *Science for Conservators* series is building on this concept. It is intended to serve as a basic primer, and an aid to the assimilation of concepts that will become embedded in the thought processes involved in decision-making by conservation professionals. The series is targeted at several groups: those about to enter a conservation programme without personally having a scientific background; conservation professionals in training and with a similar background; those who are establishing themselves in their challenging and rewarding field; professional conservators who may want to check that they are using a scientific term to mean the same thing a scientist would assume it to mean, rather than its meaning(s) in everyday conversation; and those who have studied the natural sciences or even the conservation of artefacts in another language than English, but are now working, writing reports, and publishing in English.

When reading each book, allow yourself to become completely familiar with a section and confident about its contents before moving on to the next. Do not read large portions at any one sitting. Although the series is an elementary one, you will need to take plenty of time in working all the way through it. You should not feel disheartened if your progress at times seems slow. If you do have particular difficulty with a section, ask a conservation scientist, a heritage scientist or another conservator about it. It is not worth struggling on your own; even a scientist with no knowledge of conservation can help, although one associated with a conservation programme or a museum would be even better. Very often the problem seems surprisingly simple to clear away if you can go through it with somebody else.

Worked examples and exercises have been included where they will be useful. Check your answers at the end of the book. Occasionally, some simple demonstrations are suggested to illustrate or clarify the written text. At relevant points you will also find reference tables. When scientific terms appear or are defined for the first time, they are **emphasised thus** for easy reference. A full index of such terms is included at the end of each book.

Knowledge in the conservation profession is always growing. New journals are established, and the longer-running journals can now be accessed online right back to their first volume. Conferences and events take place every year and are now preserved as recordings of the presentations and discussion sessions as well as in published conference proceedings. There are hundreds of books now written by conservation professionals for their fellow professionals, with much less need for conservators to rely on textbooks written for pure scientists. I am planning that this series will be expanded beyond the three books originally published, to cover a broad spectrum of topics relevant to the conservation of artefacts – just like the original concept of the (short) 1982 series. Further books being planned now include *Colour, Colour Measurement and Colour Change, Experimental Design and Scientific Data Analysis* and *Preventive Conservation in Practice*.

Knowledge of health and safety has changed and developed a great deal since the 1980s, of course. The conservation profession has also evaluated the effects of past treatments on the preservation, appearance and indeed survival of objects, and was beginning to do this, and publish such evaluations, in the 1980s. Many treatments that were common before and in the 1980s, and some chemicals and reagents that were then available on the bench in most conservation workspaces and every undergraduate chemistry laboratory, now appear hazardous to objects, dangerous to the user and inappropriate to use, in the third decade of the twenty-first century. Yet they played a part in the long history of a great many artefacts: it is necessary to know what might have been done to an object in the past in order to treat and sometimes stabilise it now. Also, treatment records used the terminology of their era. I have taken the decision not to exclude from the updated series the materials that are no longer used by responsible conservation professionals. They were used at the time for their unique properties and perceived efficacy, and they may have affected the properties of an object and its likelihood (or not) of surviving for another few centuries. The reasons why they fell out of regular use will however be mentioned.

Some materials derived from petrochemicals are going to be much less used in the future, as greener alternatives are developed and come into wider use. Green alternatives made locally will become cheaper than imported chemicals, or those made from petrochemicals. Using greener materials and cutting down on both waste and transport are very active topics of discussion in the field of conservation at present, and sustainability aligns with the profession's core values too.

1 What science is

Introduction

Science is a systematic and structured way of understanding the material world. Firstly, scientists aim to describe material facts in an objective manner. To help fulfil this aim, they have developed a precise language and a specialist vocabulary to describe accurately what they have learnt from their observations, and to make communication with other scientists clear and non-ambiguous. That puts them in a good position to identify what it is they don't (yet) know – and then they can formulate the research questions that they need to answer, by building outwards from current knowledge. Scientific ideas and theories are continually evolving, and being revised and honed, although by no means at an even or steady pace, as further observations and new discoveries are made.

Scientists have assimilated this language and mode of expression and use it to develop their own research further. Science enables you to understand and link phenomena which might, on the face of it, appear problematic and unconnected. Conservators, therefore, can find this precise and structured way of looking at the material world both helpful and illuminating. This book and the subsequent ones will introduce you gradually to the language of science, especially as it relates to the work of the conservator.

Observations are *facts* just as much as a measurement leads to a factual result. Professional conservators are continuously making use of observations, almost without realising it, from the moment each conservator first sees an artefact and then begins to assess it, thinking some of these questions all the time:

What is it?
What is it made of?
How can I remove it from its case or frame, or its packaging?
How can I handle it safely? (To see underneath/inside/around the back/its base ...)
Is it mechanically fragile? Then how do I support it so it can be safely moved?
How has it altered from when it was first made?
Is it actively (and chemically) deteriorating?
Given its age, is it in the condition I would expect for this type of object?
And if it is in rather poor condition, what has caused this?
Can I slow down its deterioration?

DOI: 10.4324/9781003261865-1

Apart from the first question, which can be answered through knowledge of the history of technology, experience of objects and training, all the other questions concern *materials*, the subject of this book. And these questions can be answered by evaluating observations and making sense of their implications. This book is concerned mainly with the chemistry of materials. Properties of materials, such as colour, glossy or matte appearance, strength, hand weight, whether the surface is friable or not, whether the structure of the object is rigid or can sag or break, or can be pulled out of shape like a woven textile if handled badly, all depend on an understanding of both chemistry and physics for their critical evaluation, and often on other sciences too, such as biology, materials science and mechanical engineering. But the first step towards utilising the insights offered by these sciences is to *observe* that some aspect of the object should be studied in more detail, described in a way that allows others to understand or visualise it, documented by using images and words, and possibly compare it to other similar objects, about which they already know more. The properties of materials, how scientists characterise and measure them, and how to select conservation materials that have similar properties to the materials in the object deserve a separate book – and this is not that book. Furthermore, this book will be discussing ideas that apply to conservation materials just as much as to the materials that make up objects.

A   The value of science

The insight which science can bring to you, the conservator, will provide you with greater confidence in choosing a suitable course of action when treating an object. It will help you to understand more about the historic materials you work on and also both the modern and traditional materials you might use during conservation treatment. This understanding is bound to be useful when you consider the many new materials, and processes for using existing materials, which are continually being introduced. It is important for you as a conservator to evaluate these new developments carefully yourself. It is a great advantage to be able to read the many published articles, which discuss new methods and materials, with some confidence in your own ability to understand the science behind the discussion and why it is relevant. As a conservator you are naturally cautious. Scientific understanding can help you choose sensible ways of proceeding when a problem arises, and when a choice has to be made about the next step to be taken. It can help you to organise testing of materials that are new to you – and possibly to others – more satisfactorily, to see if they can do the job you want them to do, and also to select preventative conservation measures appropriate to the object you are treating.

Not least, science can help you to be more aware of safety in the studio, workshop and laboratory, both for yourself and for the objects you work on, and both now and in the long future life you are enabling the object to have. Science is very useful for assessing risks too, which furthers safe working practices both for yourself and for those you work with.

Nevertheless, to the experienced conservator, who has gathered considerable practical knowledge and skill over the years, the scientific approach or use of the **scientific method** may sometimes appear laborious or simplistic. A conservator used to working with metal may feel able to judge intuitively how much pressure a bent object will take in order to straighten it without being damaged. A scientist, however, given the same problem, but lacking the same practical experience, might approach the task very differently. The scientist would want to identify the metal of which the object was made and would use analytical equipment to provide data about the composition of the metal. The scientist would check out what was known about the strengths of such material and, after measuring the thickness of metal, might be able to calculate the exact force required to straighten the bent object. The calculations might also give some indication of the safety margin; the extra amount of force that would cause the metal to snap. The scientist might then encounter a set-back: it is quite possible that no really useful data exists for very old and partly corroded metal that has been buried for hundreds of years. It might take considerable effort, working with replica materials, and mimicking the ways they might change during burial over a much shorter period of time, to generate the required data. Only then could the scientist calculate the force necessary, with confidence that the correct working assumptions about the object and its chemical as well as physical properties had been made. Nonetheless, with the right equipment and full information, the predetermined force could be applied in a controlled manner and the piece would be straightened.

The conservator goes through the same processes of identifying, drawing on existing knowledge, applying a controlled force, observing the response of the metal, and so, in an unconscious – and pretty fast and efficient way – is being equally scientific. The conservator in training probably does this much more consciously, perhaps against a mental check-list, to ensure that all relevant lines of thought are being followed. The main difference is that the scientist would have used an approach that relied on measurement and numbers rather than observations described in words, or instinctively acted upon. In this simple example the scientist with access to data only for uncorroded metals would not have been able to offer much help to the conservator. But it was the *observation* of corrosion that make this so obvious and the conservator was better placed by training and daily practice to make that observation.

There are many other occasions, however, where a conservator's practical judgement through sight, touch, and past experience, may be inadequate. For example, a conservator once received a metal object which was encrusted with mud. It was described as pewter, and the conservator accepted this description because of its appearance and feel. After washing off the mud, the conservator placed the object to dry in an oven at 105°C and was horrified to see that it melted. Later chemical analysis, coupled with the object's lack of provenance, established it as a modern fake made from an alloy with a very low melting point. The fear of experiencing this type of disaster must be present in the mind of every conservator. It is important to be able to judge when and how science can be of use to you – and when it is vital to think scientifically about your own observations, even when that means assessing them as insufficient to justify any action at all until you know more. Thinking like a scientist may also help you to identify assumptions that have been made by others, and then included in the documentation for an object, without any real justification for stating that certain materials are in the object.

B   Identifying materials

Everyone from very early childhood develops the ability to recognise and identify materials and objects. Amongst conservators this skill tends to become very highly developed. It is needed because to know what an object is made of is a fundamental preliminary to diagnosing its condition, identifying problems that have to be dealt with and deciding whether a treatment is called for, followed by planning a method for treatment. That treatment might be interventive – involving handling, mechanical action to the surface, or the use of chemicals to clean, reveal, and/or protect it – or it might be preventive in nature, designed to give the object a more benign and stable environment for the future, to ensure that its condition will only deteriorate slowly. Often **visual identification** seems to occur to the experienced conservator as an instinctive and almost instantaneous process. The process, however, is worth looking at in greater detail.

Pick up any object which comes immediately to hand (you may choose an object you are working on, or something in your immediate environment – a tool perhaps, or a domestic article – it won't matter what). By using your senses such as touch, sight and smell, and your experience, decide what it is made of. As a conservator, you will be repeating this process for the rest of your life! (The most challenging object you might have picked up is your mobile phone: it is clear from a glance that it is made from many materials that have unusual properties, the screen being one example.) In making your decisions pay special attention to how you arrive at your conclusions. Look at, for example, the process and reasoning behind identifying the materials in a simple and familiar object. Suppose you had picked up a chisel and identified it as having a steel blade and a wooden handle bound by a brass collar. How you did this is an interesting (though simple) exercise in the process of identification. The starting point was to recognise the function of the object. Because the blade was shiny, hard and cold to touch you knew, by comparison with past memories, that it was 'metal'. You automatically rejected the idea of the metal being silver or aluminium – it was too rigid, had the wrong shininess and did not feel the right weight for those metals. Also, from experience, you knew that steel is the best material for tools that need to cut, and therefore expected the blade to be steel. Similarly the handle looked like wood (colour, grain) and felt like wood (warm to touch, texture, weight). The yellow metal collar just had to be brass – gold, the other yellow metal, is too expensive to use on a functional object. But it might be brass plating on steel, done so that the chisel would look more like a traditional one. Observations can only take us a few steps along the road to identifying materials in detail.

With your actual example, which may have been more complex, you will have gone through a similar routine to narrow the field: first a judgement of the function and possible age of the object and then a closer look for evidence of how it was made. Comparison with your previous experience of, say, which materials were used for particular purposes in different historical periods, begins to generate expectations of what the materials are, and also what they can't be. Stylistic information may also give clues to where the object came from and when it was made.

C   Levels of identity

The process of identification, described in the previous section, used only the simplest methods. Take a look at Figure 1.1. At the level marked 'simple visual identification' there are ten broad classes of material. It is easy to classify many objects as belonging to one of these, because each class has a distinctive combination of such properties as colour, texture, density and rigidity. Your visual and tactile senses are brought to bear on the problem and you relate what you see to the properties of materials you know.

When you identify an object as belonging to one of these categories you are also saying that you expect it to show certain properties that have been observed in other objects in the same class. For instance, you might expect all objects in one category to deteriorate in much the same manner. The idea that you expect one member of a class to behave in much the same way as the others is similar to the approach adopted by scientists. By making detailed observations and measurements they are able to obtain more information about the properties of such a group. These investigations lead to more detailed classifications, and to greater understanding as to *why* the materials each behave as they do.

For many conservation problems, the level of description needs to be refined far beyond that of 'stone', 'metal' or 'wood'. (These categories are useful in

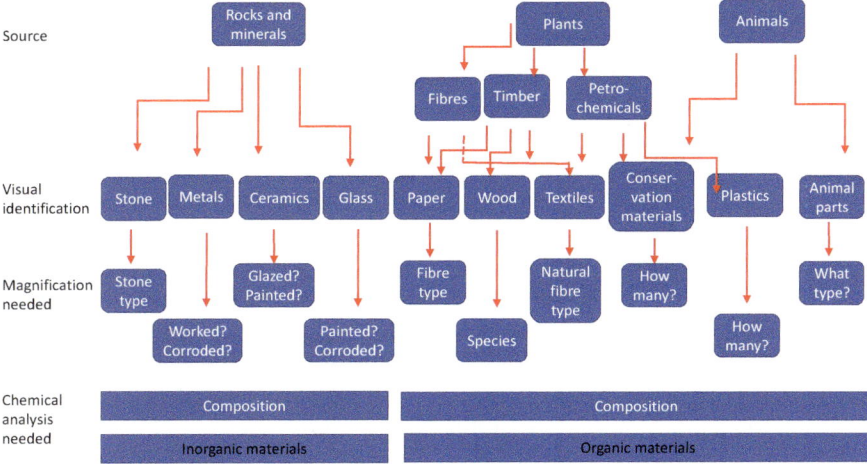

Figure 1.1 Groups of readily identifiable materials, and the levels of investigation necessary for complete identification. The two broad classes of matter (**organic** and **inorganic materials**) are related to the original sources of the materials.

themselves when it comes to preventive conservation and collections management decisions though.) The degree of refinement is dictated by the particular conservation task and the nature of the material. For instance, it may be necessary to know the exact species of wood in a piece of furniture, so that a missing piece of veneer can be replaced or so that the authenticity of the piece as a whole, as well as the damaged component, can be assessed. It has been discovered that all types of wood are basically similar in their material content, so it is not very useful to examine the chemical constituents of a sample of timber if you want to identify a particular species. What is needed is a close look at the cell structure (as a thin specimen under the microscope) which will reveal all that is necessary to identify it –always provided it is ethically acceptable and also practical to take such a sliver of wood out of the object. The fibres in different types of paper or textile that are derived from plants can be similarly recognised, by their distinctive fine structures which can be seen clearly under the microscope. **Microscopy** is shown in Figure 1.1 as the next level of investigation after simple visual identification. For conservation decision-making and for writing a medium description for a catalogue or inventory entry, **low magnification** in the range x5–50 with a stereo microscope or a low-powered digital microscope is quite sufficient for the recognition of a whole range of materials, as a class. This can distinguish many types of animal and plant product from one another. Sometimes magnifications of x100–500 may be used to identify the species of animal or plant, even if this constitutes a special skill not owned by a lot of microscope users. A research microscope with well-chosen lighting and a particular type of optical set-up might be needed for this more detailed study, as well as access to images of real samples of comparative material, naturally aged, or else a very comprehensive set of relevant modern replica samples for comparison.

At low magnifications, applied coatings, infills, coloured paint used originally or as a retouch, and adhesives, can be seen to be different from the main body of the object, although they cannot be identified in any more detail. But simply knowing whether they are present makes the conservator think differently about the object and how its bulk material and its surface appearance have survived, and it can point the way to planning what to treat, and what to remove during a treatment. For the enormous range of synthetic materials in the categories of 'conservation materials' and for 'plastics' (in the sense of objects made of semi-synthetic or synthetic polymers), microscopy is not much help: many such materials are transparent and colourless (and therefore easier to see with the microscope than with the eye alone), while others are opaque and coloured: all a microscope can really do here is to enable the user to count how many sorts of material might be present. This brings us to the more subtle level of identification labelled **chemical analysis** or **materials analysis**. For instance, you might need to know the exact nature (the chemical composition) of a corrosion product on the surface of a metal artefact in order to be sure of a safe removal procedure (if it is decided to remove it at all: if it is a compound known to be very stable, there is no requirement to remove it, to guarantee the object a long lifespan) and/or subsequent safe environmental conditions for the object (whether it is treated or not). This would involve identifying both the metal and its alteration product by chemical analysis. To recognise something as made of iron or lead or copper is, in essence, a chemical identification; the actual substance itself is being defined. On the whole, such identifications cannot be made just by looking, even under the microscope, although a few possibilities can often be excluded by this means. Gold, copper and silver have characteristic colour and lustre that make them pretty easy to recognise as consisting predominantly of one of these metals, provided they are in pristine condition. Many other metals besides silver look white/grey and can take a high polish, but would rarely be mistaken for silver. Some characteristic unique to the material must be exploited; this may be done by chemical analysis, which is mostly effected today by using instrumental analysis rather than chemical reagents. In the conservation laboratory, if only a very few options in object material are expected, chemical spot testing may be useful, but it has a weakness, in that it cannot reveal the unexpected. Chemical spot testing consists of looking *for* a specific material, or else proving it isn't there as a major component. The same applies when you need to know the exact composition of something: instrumental analysis of the right type will give an answer to the question that was asked. However, taking a tiny sample for analysis and first examining it under low magnification to remove dirt, coatings and other foreign material often gives a 'better' answer to a question about treatment than using a hand-held, non-destructive instrument close up to the object to analyse the materials at its surface. In some cases the instrument would analyse dirt on or absorbed into the surface, non-original coatings and any fingerprints that are present, and this might not give a useful answer to the question that had been asked.

Glass, for example, is easy to recognise as a class of material from its superficial properties (hardness, transparency, smoothness and all too often its propensity to have broken into pieces), even when it has been altered by long burial. Of course, not all glass is the same; a great variety of compositions is possible, since glass has been produced for thousands of years. Different forms of glass can be made from a range of basic starting materials and some added extra ones. Under a low-powered microscope the different types are not in the least characteristic or recognisable and so, should the type of glass need to be identified, a fairly detailed compositional analysis would be required, or some fairly detailed measurements made. But the microscope can help to define whether there is a problem with the glass object in question, such as flaking corrosion that has developed during burial, or scratching and abrasion from a previous attempt at replacing a missing piece, or yellowing adhesive residues where it was mended but later broken again. Alternatively, a partial analysis might be all that is necessary, say to determine the proportion of lead present, which would give information on the quality and original value of the glass, and might indicate something useful about its function when only a broken fragment has survived.

Identification may take the form of description (as with wood, paper or natural textiles), or elemental analysis (of metals or glass) or compositional analysis (of ceramics), and sometimes a combination of the two analytical methods. Often what you know about the origins or function of an object will be of great help in narrowing the field of choice in deciding what it might be made of; the more complicated (and rigorous) types of microscopical examination or materials analysis can then be applied in the light of what you know. For example, you would not expect a modern painting to be on a mahogany panel (except in a very few cases when the artist made this rather expensive choice, and said so in an interview), but even less would you ever expect an Anglo-Saxon sword-blade to be made of chrome steel, which was not available in that period. The first example would require an identification at the level of wood species (by microscopy) to be certain that is mahogany, although close observation would have shown that it was not a softwood panel. The second example would require materials analysis and a research microscope to identify the composition of the metal blade and the kind of metallurgical processes used to create it.

Having looked at the means of making increasingly specific and detailed identifications and analyses of materials, the reader should look again at Figure 1.1 and in particular at the two large rectangles marked **inorganic** and **organic**. You will likely already be familiar with the idea that stone, metal, ceramics and glass are all ultimately derived from rocks and/or minerals and are termed inorganic. The idea that wood, paper, and many textiles are derived directly from plants, while wool, silk, leather, fur, horn and bones are all animal products will also be straightforward enough. Referring again to Figure 1.1, you will see that they all appear within the rectangle marked organic. What may well appear as more surprising, however, is that many **synthetic materials,** many **conservation materials** (e.g. all synthetic polymers, solvents, Paraloid® resins, polythene, etc.), some of which are made from chemicals derived from animal and plant products, while others are **petro-chemicals** and thus based on refining oil and gas to make fuels, are also organic materials. (Do not forget that many substances, although looking deceptively like inorganic materials are, of course, derived from animals or plants. Natural gas and fuel oil are both derived from organic materials, namely plants and animals that have been altered chemically over geological time.) There are, too, inorganic materials that may be either natural or synthetic in origin. For example, the pigment vermilion can occur naturally as the mineral cinnabar and can also be manufactured from mercury and sulfur. The two forms, provided they are pure, are chemically identical, but in this case they can often be distinguished by microscopy.

The terms organic and inorganic distinguish two groups of material from different sources. This division by source is shown at the top of Figure 1.1. You might expect that there would be an equally obvious distinction to be discovered by the investigation of their composition. This turns out to be the case. The words organic and inorganic as chemical descriptions will start to have greater meaning as your appreciation of material in chemical terms increases. In terms of collections care, organic materials are far more readily, rapidly and more severely responsive to their environment than many inorganic materials.

D   The use of instruments and scientific language

The fact that your own methodical approach to your work is 'scientific' – based on observations followed by deductions from them – may be obscured for you by an idea that scientists are different in some way from other people. People without scientific training naturally notice that 'science' involves the use of instruments that in appearance resemble a 'black box' or at least a grey or cream-coloured one with a laptop running it, and an apparently complex language. You may well feel, quite subconsciously, that 'scientists' are much more intelligent than you are, or that their brains work in a different way, or that they are operating within a kind of intellectual 'closed shop'. None of these feelings represents any sort of truth. The use of highly specific instruments comes about from the need to make observations on a very minute level; the use of 'off-putting' language from the need to describe what has been observed or discovered, in a concise manner, using a single word or phrase that has a previously-defined meaning. In the previous section, more complex ways of identifying materials were suggested and these tended to imply the use of instruments or else a knowledge of materials analysis. It was suggested that you might, for example, use a microscope to extend your powers of vision when identifying paper or wood.

The use of instruments is obviously not restricted to identification: they can provide a **measurement** of some property of a material that can be replicated by other users of the same instrument. If you wished to maintain correct storage conditions for an object, you would need, amongst other things, to monitor the temperature of its environment to provide greater accuracy than merely feeling whether the room is warm or cool, by obtaining a measurement of the room temperature in degrees. Simple measurements such as weighing an object are very useful for planning its safe handling, packing for transport from the conservation laboratory or studio, and storage. Simple judgements like assessing whether an object's surface is cohesive and strong, or flaking and powdery, hence very easily damaged by necessary handling during each treatment, lie at the junction between observation and measurement. The conservator might very gently touch the surface with a swab stick before even handling the object, to assess the fragility of the surface. The museum scientist – or the conservator – might use specialist lights or imaging techniques the better to observe the surface and what lies just beneath it.

Because all scientific thought and activity is based on making detailed observations, scientists have needed to develop and use instruments of varying complexity as a means of measuring and then interpreting what they have observed. Instruments often relay the information they are designed to detect in terms of numbers. Examples are a measuring tape marked off in cm and mm, or a pH meter which indicates acidity or alkalinity in terms of a scale ranging from 1 (highly acidic) to 14 (highly alkaline).

It will be quite obvious to you that your work as a conservator can depend on the correct use of instruments (of many different kinds and for differing purposes) and on your ability to use them both appropriately, and safely for the object. The information which an instrument may offer is usually limited in kind although instruments are normally able to detect and quantify far beyond the ability of unaided human senses. It is also easier and almost always more accurate to make a before-and-after comparison using an instrument. It is partly for this reason that much of the data instruments give can appear rather abstract or obscure, particularly as many of the phenomena described by scientists are only detectable with the aid of instruments. To describe things that are not obviously a part of the everyday world of the senses, new words have been created and these have been incorporated into **scientific language**.

Every new discovery (not only in the field of science) has meant that new words have had to be created or old words given more specific meanings, to describe what was previously unknown. The language scientists use may at first appear off-putting. However, it has a regularity and pattern which, once several fundamental scientific ideas have been understood, makes it far more consistent and comprehensible than might at first appear. The scientific language, like the instruments you use, is aimed at providing a precise and accurate means of describing the phenomena investigated by scientists. This means that as you read the books in this series, you will find that certain words, used freely within normal conversation (for example, words like buffer, reaction, stress) have a very specific meaning within a scientific context. Other words (such as carbon dioxide) will tell you something about the substance itself, and hence its properties, once you have begun to understand a little about chemistry. Others still (such as esters, isotopes and polymerisation) are found in the language of science alone. Along with the new words there are also **symbolic representations**, and these are especially prevalent in chemistry relevant to conservation treatments, and in physics-based methods for analysing and dating materials. The chemistry symbols are often combined to form equations, designed as shorthand notation to describe chemical processes. It is hoped that by the end of the *Science for Conservators* series your understanding of the language and vocabulary of science will be sufficient for you to read most technical articles on conservation subjects with some understanding of the scientific principles involved.

E   Observations and theories

It is a common misconception that 'science' represents incontrovertible truth. While science is concerned to represent facts on the basis of consistent observation as objectively as possible, the scientist has to look for a way of describing what has been observed. Because scientists are always aware that their descriptions of phenomena are often only visualisations of what cannot be seen but must exist in reality, they often prefer when describing something to refer to their description as a (conceptual) **model** for understanding it. This word reminds one that science is not a series of static or absolute statements about the material world, but rather a *framework* by which to understand it. It is a continually evolving process that is constantly being revised and developed further as more observations are made. And if new observations don't make any sense in terms of the current model, and the observations are sensible ones and can be made repeatedly by other scientists who are investigating the same set of circumstances, then it is the model that is in need of improvement.

The scientific way of thinking and acting is, at root, simply an extension of natural common sense, curiosity and intelligence. It relies on our predilection for observing situations and occurrences and our ability to detect patterns and connections within them; the new insights might lead to a **hypothesis** that when something is done, and in such a way, then a certain result will follow. Consistent observation of a particular pattern of events may lead the observer to devise the beginnings of a **theory** (a statement of what is likely to be true under certain conditions, arrived at through detailed observation and **experiment**) to explain the consistency, and then also to define the limited circumstances when it will be true. A well-developed theory also discusses *why* the result follows, as well as *when* it follows. Such a theory may then be tested by experiment(s) or by further observations. If observations and experiments suggest that a particular occurrence is always, without exception, accompanied by a particular pattern of consequences, this may be stated as a 'law'. A **scientific law** does not dictate to nature what will happen, on the contrary it says that 'because this has always been observed to be the case, it probably always will be'. Thus it is, in contrast to the meaning of 'theory' that is used in everyday speech, not as far-reaching or of the same importance as a theory which describes the underlying causes of the phenomenon, and also makes predictions that can be checked out. Scientists are always trying to disprove theories by testing them beyond the set of circumstances where the theory is known to hold good, from many experiments and observations. Simply observing that the theory is still holding good is not what they do. A scientific theory is in effect a 'working theory' in the everyday sense: its principles tick over in the background of one's mind, it is updated from time to time, and once in a while it will be found to apply to a more limited set of circumstances than one had first assumed it would. And then the model can be narrowed down, to that set of circumstances, and a new research question can be asked: what is going on outside these circumstances? The whole process and scientific writing, peer reviewing and checking of what others have written, and publishing, is designed to capture existing knowledge so that new and harder questions can be asked, and then answered.

The relationship of observation to hypothesis, experiment and theory, can be illustrated using the example of the fading of textiles caused by light. An observant person might see that some curtains in an historic interior had faded quite badly and that the cloth was falling apart. By making further simple observations this person notices, too, that other window curtains through the historic house fade and deteriorate, and that carpets and upholstery also located near the windows fade rapidly, although tapestries and tablecloths further away from them are not so badly affected, though all of these objects have been in the same place within the room for many decades. What do the faded textiles have in common? The observations are sufficient to suggest an idea (hypothesis) that there is a connection between high exposure to daylight and the fading and deterioration of textiles. The observed changes cannot be due to handling, as a once frequently-used tablecloth, for example, has not suffered so badly. It cannot be (only) the difference in temperature between the window and the middle of the room because a chair in front of the window has faded but one right next to it in the shadow has hardly changed, and indeed the side of the chair farthest from the window is less faded. The idea that fading is related to light falling on the material is only a hypothesis (a surmised truth on which to base further reasoning) until the relationship has been demonstrated to hold true. It could be proved by making a large number of observations to confirm that where textiles are kept in light they always fade but when they are stored in the dark they never do. Alternatively, it could be confirmed by a controlled experiment in which a comparable textile is deliberately placed partly in light and partly in shadow (or covered up) and the different reactions observed over time.

The observer may also develop more complicated hypotheses – that the amount of both colour loss and loss of strength in the textile fibres depends on the quantity of light that has fallen on the material, or that light of one colour causes more damage than another. Another hypothesis based on observation might be that silk is soon damaged by light, and always loses strength faster than woollen fabric in the same circumstances. These hypotheses are best verified by **controlled experiments** – which are examples of the **scientific process** – in which the variables such as light intensity, duration of exposure and colour change can be accurately measured. Varying just one thing – the amount of light – makes for an easier experiment because only one thing needs to be measured, and it seems at first thought that it must lead to a clear-cut conclusion, but it is still necessary to think about whether it is really just one thing that is being changed. For example, using a sunny south-facing window (assuming one lives in the northern hemisphere: in regions south of the equator this applies to a north-facing window) would guarantee a lot of light in summer, but probably the sun-exposed textile samples would often get hotter than the ones kept in a box under the window, in the shade. It might be better to wrap the textile samples that are not to be exposed to sunlight in something very opaque, and leave them in the sunlight too. In other words, **planning an experiment** is a skill that requires a lot of lateral thinking.

To aid understanding and in an attempt to explain these observations to others, the experimenter may develop a **theory** of the fading of textiles by light. This theory will combine the observations, the results of the experiments, and hypotheses about the nature of light or the chemistry and properties of the textiles that lead to the phenomenon of colour loss, which, although necessary to the theory, cannot be proved at that time. If the experiment is then carefully reported in a publication, along with the assumptions made at the time, it is easier for the next experimenter to take the next step without repeating the past step, then work out an answer to his/her own research question, and thus to build on past knowledge and increase what we can all know now. The value of a theory is that it can be used to predict how a particular substance will behave in a particular situation. However, the only way to know what will happen is to do the experiment and make the observations. Thomas Huxley referred to 'The great tragedy of Science – the slaying of a beautiful hypothesis by an ugly fact'.

Here it must be emphasised that an 'experiment' need not involve numbers or obvious measurement techniques at all. At the planning stage, it might be a **thought experiment**: a mental exercise that involves thinking about what would happen if the same conservation treatment was done to, for example, a new and pristine piece of metal rather than the corroded, fractured and long-buried metal object on the bench, or if the treatment was carried out at the below-zero temperatures at the site where it was newly excavated, rather than in a climate-controlled conservation workspace. All these thoughts and all the observations of the object and the materials that might be used to clean or treat or protect it after cleaning are just as valid for the situation as generating numbers and plotting them on a graph to illustrate a hypothesis, even though they are **empirical**, meaning based on observation and rooted in common sense, rather than feeling inherently 'scientific'.

It is important to remember that conservation professionals take a very long view of the lifetime of cultural heritage. While nothing can last forever, a great majority of objects in collections, and buildings too, are valued as cultural heritage for immensely longer than their 'working life'. In a real sense, the conservation profession takes over the task of ensuring survival of an object at the moment when its original maker would see the end of its working life, the time when a decrease in optimum appearance, strength and usefulness has occurred, and a new and/or improved one would fulfil the original function just as well if not better. For more modern and contemporary objects, this means that industrial research into materials, highly science-based and well-funded though it may be in comparison to museum-based research, is not always useful for understanding the very long after-life of the materials making up an object, and preserving it for centuries rather than decades. The conservator is nearly always dealing with an ageing and altered object, and will find that published information dealing with its constituent materials is often inadequate. That is why observation, then rationalising the observations, employing a scientific approach to decide what else needs to be known and how it can be found out in practical terms, is necessary for making the right decision even when knowledge is less complete than one would like it to be. At its most rigorous, and when it involves numbers derived from measurements and comparison of different objects through calculations, this way of thinking is known as the **Bayesian statistical approach**. It's what conservators do in their heads, and almost without knowing that they are doing it.

F   Measurement and accuracy in practice

Through reading the previous sections it will have become increasingly clear to you how much science relies on making disciplined and accurate observations, done in the same way each time, described sufficiently well that someone else could come along to make the same kind of measurement independently, and get the same result for the material, and followed by a process of thought that leads to interpretation. Many scientific observations are based on repeated measurements, although some require the use of sophisticated and often expensive instruments. Generally speaking, these specialised instruments need trained personnel to operate them, to assess the appropriateness of using that particular instrument in any application, and to interpret what the numerical results mean. It's a skill that can take a full-time scientist 6 months to become proficient in, just for one type of instrument, followed by 5–10 years regular use before reaching 'expert' status. Conservation studios and laboratories are not usually equipped with the 'standard' and rather small number of types of analytical equipment to be found in a museum conservation science laboratory, and so conservators will probably only have regular access to them through consultation with others, during their training programme, or when carrying out an extended research project. This is probably no great disadvantage because often much more modest techniques can adequately solve many practical conservation problems, in a shorter time than it takes to become proficient in anything more than the art of taking a good sample or identifying a good spot for a measurement that someone else might be making. But whether 'high technology' science or simple methods are used, there is always a need to understand and use **sound experimental techniques**.

'Sound experimental technique' describes a systematic and well-informed approach to the factors which may affect any practical work being undertaken. In a conservation workspace it could, for example, be measuring out the correct weights of substances in order to ensure that they form a solution of the right strength for a particular job. It could mean obtaining an accurate reading using a pH meter (see the book on *Cleaning* for this). It might involve conducting some trials in a manner that will produce helpful and reliable results, such as testing whether the colours on a textile will run when it is washed in water, or when solvent is spotted on locally to reduce the visual impact of or to remove a stain on a sheet of paper.

There are, of course, many instances where it will be difficult for you to know exhaustively all the variables that may affect your practical work. However, just as you would guard against accident by ensuring that an object is placed in a safe position on your table, so common sense and an understanding of science will show that there are several fundamental and often quite straightforward factors to be considered. It will gradually become less difficult for you to judge what these are likely to be in a given situation as your understanding of basic science develops. Once you are able to judge the variables likely to affect the results of your work, and when you are able to understand why they do, you will then have the means to find ways of controlling them.

Measuring relative humidity

The measurement of 'humidity' has been chosen to illustrate this systematic approach to practical work, because it will be familiar to most conservation students and conservators and because the factors affecting its measurement are quite simple to control.

Ask yourself the following questions:

What is 'humidity'?
Why do I need to know about humidity?
What causes changes in humidity?
What does the special term 'relative humidity' mean? It is usually shortened to 'RH'.
How is RH measured: directly, or are there any calculations involved?
How do the measuring instruments work?
How accurate do the measurements have to be?
What affects the accuracy of the measurements?
How do the inaccuracies show up?
How can inaccuracies be prevented or kept to a minimum?

All these questions are answered to some extent below, though not necessarily in the order they were asked.

It has been found, through long observation, that the majority of objects conservators work on are affected by the amount of water in the atmosphere in one way or another. Most objects made of organic materials have a measurable moisture content, because they are acclimatised to their immediate environment, and the air both indoors and outdoors pretty well always contains some water. Condensation of water on the object can induce chemical reaction, and water vapour can move from the object into the air, or the other way round. In damp conditions metal objects may corrode or even form different types of corrosion product than they would in dry conditions, and mould will grow on organic materials such as paper or leather. When the air is excessively dry, furniture made from wood may crack or veneer may lift off. This is an example of damage caused by actual changes in moisture content: materials expand as this rises and contract as it falls, and in a dry environment. An object that contains several different materials which each respond differently to changes can warp and the materials will then separate, causing considerable damage. This makes it important to be able to control moisture content through humidity control, and the first step in doing that is to be able to measure something. The next step is to measure 'it' regularly to see if and how it changes.

'The humidity' is the amount of water retained as vapour in air. It is expressed as the weight of water in a given volume of air, at the time of measurement. This measurement is called the **absolute humidity** and is usually given as weight (as the number of grams of water vapour) in a volume of air (such as a cubic metre of air, a volume that measures 1 metre wide x 1 metre high x 1 metre deep). This value is then written as a number followed by the units used to express it, as g/m^3 or gm^{-3}, which is read aloud as 'grams per cubic metre' in both cases. The notation 'gm^{-3}' is the one used by scientists and is a better notation to use in written reports and formal publications. The highest value possible for the absolute humidity is the maximum amount of water that a cubic metre of air can contain, at a given temperature, and this value would be calculated by careful and repeated experiment.

For objects, however, it is *relative* humidity that is important. Air at two different temperatures may have the same moisture content and yet have very different effects on moisture-sensitive objects. Air at 30°C containing $10g/m^3$ of water causes an object to dry out, yet if this air is cooled to 10°C condensation could occur on the object's surface, because the cool air cannot retain all that water vapour.

Relative humidity (RH), as the name implies, is an expression of one moisture content measurement relative to another. The two measurements are:

- the actual amount of water vapour in a given volume of air at a particular temperature; and
- the maximum amount of water that could be associated with the same volume of air at the same temperature.

The actual amount is expressed as a percentage of the maximum amount.

At 30°C the maximum amount of water that can be associated as vapour in a cubic metre of air is 17g/m³. Suppose the actual amount of water present is only 10g/m³. We need to express 10 as a percentage of 17 to get a figure for the RH. To do this we divide 10 by 17 and multiply by 100.

$$RH = 10/17 \times 100 = 59\%$$

The simplest methods of measuring RH rely on the expansion and contraction of a moisture-sensitive material as the RH rises and falls. But this is not a direct measurement of RH: it has to be **calibrated**. In the mid and later twentieth century, hygrometers containing elements of paper or hair were very commonly used instruments for measuring RH. The indicator needle moved along a scale in response to the movement of a paper strip or bundle of hairs that would expand and contract as the moisture content in the surrounding area affected the moisture content of the paper or hairs. That gave a 'spot reading', an indication of the RH at the moment the reading was observed or written down.

A more useful instrument for museums, the recording thermohygrograph which used a bundle of hairs, was once regularly used in every country to keep a record of RH and temperature over a period of time, usually one week or one month. The bundle of hairs contracted as the RH fell and by a series of levers pulled a pen down on the chart which was on a drum that rotated slowly, usually by a hand-wound clockwork mechanism. The pen moved up the chart as the hairs expanded with rising RH. However, both the paper hygrometers and the recording hygrograph gave increasingly inaccurate readings after they had left alone for a while, and had to be adjusted before they would read correctly again. This **calibration** required a measurement of RH by another means that was known to be consistently accurate, to predict the amount of adjustment required. The slow drift away from accuracy was caused by the moisture-sensitive element losing its elasticity and becoming stretched, and so failing to return to its original tautness after expansion, and/or by simple clogging with dust. Used on their own for years, these instruments were useless because the drift from accuracy could not be worked out and compensated for later.

A psychrometer can be used for calibration of kit that measures RH, the most basic being the **sling psychrometer** or **whirling psychrometer** which is still a very reliable and precise means of measuring RH. It relies on the cooling effect observed when water evaporates in a flow of air. The drier the air, the faster water will evaporate into it and the greater the cooling effect will be on the surface it is leaving. In a sling psychrometer two identical thermometers are fixed side by side. The bulb of one of them is surrounded by a fabric sleeve that is well moistened with distilled water. This is called the wet bulb, the other being the dry bulb. The evaporation of the water from the wet bulb is accelerated by passing a current of air over it, achieved by whirling the instrument for a few minutes, whilst holding its handle and whirling it around without whacking one's wrist or anything else nearby, and then slowing it to a stop and rapidly noting the temperatures indicated by both thermometers. The drier the air, the lower the wet bulb temperature will be compared with the dry bulb. Battery-powered RH meters blow air over the wet bulb with a fan, to achieve the same effect, and they also need to be run until the reading stabilises, the fan having powered up enough to blow the air at the required speed.

After reading the two thermometers the wet bulb temperature is subtracted from the dry bulb temperature to give what is called the depression of the wet bulb, the amount it reads lower than the dry bulb. Using this figure and the dry bulb tempera- ture the RH can be looked up (Table 1.1). The column on the left is the dry bulb temperature and the row across the top is the difference between the wet and dry bulb temperatures. The RH is read by following along the line from the dry bulb temperature until the column for the appropriate temperature difference is reached.

If the dry bulb temperature is 22°C and the wet bulb temperature is 17.5°C the difference between these two is 4.5°C. It can be seen that the RH corresponding to a dry bulb temperature of 22°C and a depression of the wet bulb of 4.5°C is 64%.

Table 1.1 Portion of the psychrometric conversion chart.

Dry Bulb (C)	Depression of the wet bulb (•C)														
	0	½	1	1½	2	2½	3	3½	4	4½	5	5½	6	6½	7
40	100	97	94	91	88	85	82	80	77	74	72	69	67	64	62
39	100	97	94	91	88	85	82	79	77	74	71	69	66	64	61
38	100	97	94	91	88	85	82	79	76	74	71	68	66	63	61
37	100	97	94	91	87	85	82	79	76	73	70	68	65	63	60
36	100	97	94	90	87	84	81	78	76	73	70	67	65	62	60
35	100	97	93	90	87	84	81	78	75	72	70	67	64	61	59
34	100	97	93	90	87	84	a.1	78	75	72	69	66	64	61	58
33	1OO	97	93	90	87	83	80	77	74	71	69	66	63	60	58
32	100	97	93	90	86	83	80	77	74	71	68	65	62	60	5
31	100	96	93	90	86	83	80	77	73	70	67	64	62	59	56
30	100	96	93	89	86	83	79	76	73	70	67	64	61	58	55
29	100	96	93	89	86	82	79	76	72	69	66	63	60	57	54
28	100	96	93	89	86	82	79	75	72	69	65	62	59	56	53
27	100	96	92	89	85	82	78	75	71	68	65	62	59	55	52
26	100	96	92	88	85	81	78	74	71	67	64	61	58	55	51
25	100	96	92	88	84	81	77	74	70	67	63	60	57	54	50
24½	100	96	92	88	84	81	77	74	70	66	63	60	57	53	50
24	100	96	92	88	84	80	77	73	69	66	62	59	56	52	49
23½	100	96	92	88	84	80	77	73	69	65	62	59	56	52	49
23	100	96	92	88	84	80	76	72	69	65	62	58	55	61	48
22½	100	96	92	87	83	80	76	72	68	64	61	58	55	51	47
22	100	96	92	87	83	79	76	72	68	64	61	57	54	60	47
21½	100	96	91	87	83	79	76	71	67	63	60	57	53	50	46
21	100	96	91	87	83	79	75	71	67	63	60	56	52	49	46
20½	100	96	91	87	83	79	75	71	67	62	59	56	51	49	45
20	100	96	91	87	83	78	74	70	66	62	59	55	61	48	44
19½	100	96	91	87	82	78	74	70	66	61	58	55	51	47	44
19	00	95	91	86	82	78	74	70	65	61	58	54	50	46	43

To use a psychrometer-type instrument correctly, that is, to obtain accurate experimental information from it, certain precautions must be taken. The most common inaccuracy is to have too high a wet bulb reading. This gives too high a value for the RH. If the dry bulb temperature is 22°C and the wet bulb reads 16°C instead of 15°C then the RH will be calculated too high, as 54% instead of 47%.

Types of carelessness in use that can lead to high wet bulb readings include:

- not whirling for long enough to allow the air to flow over the wet bulb before reading the thermometer (2 minutes minimum was recommended for the type used in museums);
- too long a pause after whirling before reading the wet bulb thermometer;
- breathing over either thermometer or putting warm hands on them while taking the reading;
- allowing the wick to get dirty, or stained from non-deionised water or repeated wetting and drying, or not using distilled water, any of which would reduce the amount of water evaporating off the wet bulb.

Thus, in order to obtain a reliable measurement of RH it is important to use the psychrometer correctly, and this means understanding how it works and *why* it is used the way it is used. This is a vital principle in any experimental technique. Some of these considerations apply just as much to the hand-held, battery-powered RH meter too.

To improve the precision of the measurement the thermometers could be read to the nearest 0.25°C rather than the nearest 0.5°C as shown on the psychrometric chart (Table 1.1). It can be seen from the table that there are quite large jumps in RH between one column and the next. For example, if the dry bulb temperature is 19.5°C and the wet bulb reads 15.25°C then the difference in temperature is 4.25°C. The RH reading for a temperature difference of 4°C is 66% and for 4.5°C is 61%. A value of RH halfway between 66% and 61%, that is 63.5%, would be more useful. If the thermometers in the psychrometer can only be read to the nearest 0.5°C then the precision of the measurement will be limited to within +/-2.5% RH.

A precision of within +/-2.5% of the true value may be acceptable. Fluctuations in RH of this order may not have any serious effect on wooden furniture, for example, especially if they happen so fast that the moisture content of the wood cannot respond to the changing moisture content of the surrounding air, before it goes back to the initial value. However, if conditions for the storage of metal objects have to be maintained below 40% RH to avoid corrosion, a reading of 35% is definitely safe, but one of 39% may not be. Hence cheap digital RH meters, like neglected hair hygrometers that give values that are too low or too high by as much as 10% RH, are quite unacceptable to use for monitoring such storage conditions.

Accuracy (goodness of match to the real value of what is being measured) and **precision** (scatter around the value that is obtained, by several measurements that are then averaged) of measurement is important in a great deal of practical conservation work, but so too is knowledge of when accuracy is needed. A cheap digital RH meter may give the same value in successive minutes of measurement of the museum environment, which means it has quite good precision and repeatability. But that repeated reading might not be accurate, if the meter cannot be calibrated at all, or if it had not been calibrated or compared to a better instrument for months or years. As another example, there are certain cleaning solutions that you must use at an exact concentration, but there may be others in which the concentration is not so critical. If you were washing a delicate piece of historic costume it would be necessary to weigh accurately the ingredients for the cleaning solution. It is important that no residue remains in the textile and rinsing must be kept to the minimum, as the more the object is handled, the greater is the risk of damage. However, it is not so important to measure precisely the amount of detergent that you add to the water when you clean your lab coat in a domestic washing machine. You will be better able to make this sort of decision, and to see the reason behind it, when you have acquired a basic knowledge of chemistry.

2   Beginning chemistry

Introduction

The first chapter introduced you in a general way to the structured way of thinking that science uses, showing you that in many instances this is much the same as the normal, common-sense approach used in conservation work. This chapter introduces molecular theory at a simplified level and shows how it can be used to interpret several commonplace phenomena.

A   Chemical names

The words that are used to describe a material or a substance can give a great deal of information about its nature, or they may tell us very little. The name 'Jane Smith' may identify one person within a group, the word 'conservator' may describe how she spends her time during working hours, but we need words like animal, mammal, human and protein to give us more and more specific information about the structure and composition of Jane Smith, conservator.

If we say 'stone', this word refers to a whole class of materials (rocks) that have recognisable properties of hardness and density and have a common provenance – they occur as rocky outcrops in landscape. The name 'marble' defines a narrower group within this class, defined by some of the geological processes that lead to the formation of marble, but there is nothing in the name that helps predict a relationship with any other group of rocks. However, the name 'calcium carbonate' describes marble in such a way that someone with a little knowledge of chemistry but no practical knowledge of marble or limestone could make reasonable predictions about the way both would react to acid cleaning or be affected by acidic air pollution. 'Calcium carbonate' can equally be used to mean limestone, a soft rock with very different carving properties from marble, and a sufficiently different appearance that a glance can distinguish them, or to describe many chalk cliffs and mountains. In some countries however, the white chalky rock is dolomite, which is a combination of magnesium and calcium carbonates, and only one part of the name applies to all chalky white cliffs worldwide.

Many different kinds of name are used by conservators and by scientists. There are, for example, names that describe classes of materials. Amongst these will be

DOI: 10.4324/9781003261865-2

those that describe (only) the function of a material, such as 'thinner' or 'diluent'. Although people who use paints, coatings and solvents will know what 'thinners' do, the name doesn't indicate at all what thinners are made of, or even whether all thinners are the same, or whether a given bottle of thinner is suitable for all possible types of paint. It doesn't enable users to look up toxicity or flammability data or to record their work with such accuracy that someone else could exactly repeat the procedure on another occasion. Other 'function' names are gelling agent, and enzyme, for example.

Another type of class name is the **commercial name** of a product, which may be a **registered trademark**. An example is Araldite®, which describes a range of products whose composition is exactly known by the manufacturer (at least, known for the time period when it is still in production), marketed with numerical codes that define each product. However, the user may only know that Araldite® is often some kind of 'epoxy resin' that comes in two components to be mixed together at room temperature for use, with no indication of why or how it hardens or what vapours may be given off, and the Araldite® product range might even include other types of resins as well as epoxy type. The manufacturer's advertising might emphasise that 'Araldite® sticks things tightly' although that only means that it can form a good adhesive bond between two surfaces of the right sort. Commercial names are also inadequate because the manufacturer may change the composition (to provide a safer or greener product, for example, or else one cheaper to produce, or more suitable for export to another country with different health and safety regulations) without changing the name or the code. If a product is sold for decades with the same commercial name, it is more likely to have changed or been 'improved' than to be the same as when first sold, in fact. Older publications on its use in conservation might be referring to an older formulation for the product, not the current one.

Commercial names can fall out of working knowledge pretty fast, when the product goes off the market and a different one is introduced. It's important for future conservators that the full product code of the Araldite® is written down in treatment reports, ideally in the style 'Araldite® 123 2-component epoxy resin weighed and mixed as directed'. In a publication, this level of detail is always required, and the name of the supplier should be given too. A conservator in another country might need to check on the company website whether it is called Araldite®123 on another continent, or whether it is sold under a different name. Even more critically, the conservator of a later generation, rubbing down the gap fill made of an Araldite® product in an object, might need to work out in advance whether the dust from the process posed any health risks that were unknown to the person who used it the first time.

Pure substances are given **specific chemical names** by scientists and others who use them frequently. Amongst this group are some simple names such as 'acetone'. These names refer only to one pure compound and once it has been generally accepted that the name is only ever used to describe this one material with a particular composition and structure, it can be used, without confusion or danger, in giving recipes for conservation treatment or in discussing chemical reactions. However, there is nothing in that name which indicates what the structure or properties might be; they are the chemical equivalent of a person's first name. For this reason they are often called **common names** or **trivial names**.

In most countries including the UK, EU countries, USA and Canada, a **globally harmonised system (GHS) for labelling chemicals** has been adopted. Since even the more standard chemicals can usually be purchased at better than 99% purity, as well as more expensively for analytical purposes at a purity that might be better than 99.99%, the **safety data sheet (SDS)** that a manufacturer must provide on request can be usefully applied to the same chemical in stock, but provided by another manufacturer. The SDSs provided in North America have long been known in the UK as **materials safety data sheets (MSDSs)**. They list the main hazards and health risks posed by the material, including whether it is flammable, harmful or toxic, whether it exists at room temperature as a vapour or very fine powder that could be breathed in, is a solid that could be absorbed through eating it due to spills on the hands, or is a liquid that could cause eye damage if sprayed carelessly. The information sheets for chemicals that are liquids at room temperature also give figures for boiling point, density, and other physical properties that give the reader with scientific knowledge some very useful information about how to handle the liquid safely. They use standard **risk phrases** that can readily be translated into other languages. Commercial products, of course, need not consist of a single chemical compound, and in fact very rarely do. Acrylic paints, for example, can easily include six to ten different distinct chemicals before the coloured pigments are even included. Manufacturers are obliged to list all the major components in terms of volume in their SDSs, with information on the risks posed by each, but they do not need to report on hazards associated with chemicals present in only trace amounts, less than one percent.

There is a second group of specific names which have been produced by following agreed rules of naming. This is managed by the **International Union of Pure and Applied Chemistry (IUPAC)**[1], and for this reason these are called **systematic names**. Examples are 2-propanol (also called propal-2-ol) and sodium chloride (common salt is its common name). These names contain information about the component parts and in some cases information about the structure of pure substances. Scientists often use a mixture of trivial and systematic names in publications as well as in everyday conversation, so that they can avoid some of the tongue-twisters that rigid adherence to the rules would produce. This habit results in the loss of some structural information but as long as the name is specific to one substance there is no problem.

B   Elements and compounds

If you look at some of these commonly used specific chemical names you will notice that some are single words and some are combinations of these words. For instance, the atmosphere in a city will contain a number of gases, including: oxygen, nitrogen, argon, carbon dioxide, carbon monoxide, nitrous oxides and sulfur dioxide. It might include small amounts of hydrogen sulfide as well, and in the past

---

1   http://www.chem.uiuc.edu/GenChemReferences/nomenclature_rules.html

industrial cities certainly had this gas in the air, at dangerous levels for people and for materials. In this example the single words (such as oxygen) all name **elements**. The two-word names refer to **compounds** that are made from two elements combined in some way. The term 'nitrous oxides' in this list was used to refer to a class of compounds, not just a single possible combination, because several can be found in city air. In the cases given you can see the element names in the individual parts of the combination, though with slight changes (nitrogen to nitrous, oxygen to oxide). The compound names tell us which chemical elements have combined to form the chemical compounds. Thus carbon and oxygen are the names of elements, and carbon monoxide and carbon dioxide are the names of compounds formed by different combinations of the two. (The prefixes mono- and di- mean one and two respectively, and their use in the names suggest that carbon dioxide contains twice as much oxygen as carbon monoxide. This is intuitively obvious to speakers of some languages, but less obvious to those who learnt English as their first language.)

Although there are thousands of distinct compounds these can all be generated from a relatively small number of different kinds of elements, or small groups of elements joined together in various combinations. This is the basis of **molecular theory**.

Altogether there are 92 elements that occur naturally on our planet, mostly in a non-radioactive form, while 26 more have been made artificially in nuclear reactors, though not all of them last for long once made. Each different element is given its own name, which will be different in languages other than English, and a symbol consisting of one or two letters, such as O for oxygen and Ca for calcium, with the same symbol used in all languages. The symbol is used for simplicity's sake as a shorthand for the full name, and it is very useful when describing compounds.

In the materials that a conservator is likely to handle, the names and symbols of fewer than 40 or so elements need to be recognised on a regular basis. Some of the most common ones are listed in Table 2.1. You can see several familiar names in this list; many of the most common elements have been well known for centuries.

Table 2.1 The standard symbols for some elements that will be discussed in this book.

Element	Symbol for the element
aluminium (aluminum is the accepted spelling in the USA)	Al
calcium	Ca
carbon	C
chlorine	Cl
copper	Cu (from the Latin *cuprum*)
gold	Au (from the Latin *aurum*)
hydrogen	H
iron	Fe (from the Latin *ferrum*)
nitrogen	N
phosphorus (phosphorous is the accepted spelling in the USA)	P
silicon	Si
sodium	Na (from the Latin *natrium*)
sulfur (sulfur is the correct USA and IUPAC spelling, but UK English still sometimes uses sulphur and this spelling was once more common)	S

Usually the symbol is formed from the first letter of the name and sometimes one other letter. This second letter is necessary because there are only 26 letters in the English alphabet. Some elements that have been known for a long time have symbols that are derived from the Latin names used by alchemists since the medieval period, e.g. Au for gold (*aurum*).

You know from practical experience that carbon black pigment, made from pure carbon, is, obviously enough, black in colour. Among metals, aluminium is matte 'white' in appearance, but clean and polished silver has a glossy 'white' appearance, while polished copper is called 'red'. You also know (at least, after thinking for a moment) that the oxygen you breathe in and the carbon dioxide you breathe out must be colourless transparent gases, because they are as 'invisible' as air is. Consequently the compound carbon dioxide must be more than just a finely divided mixture of the black solid and the colourless gas, for the blackness and solidity are totally absent in the compound. Similarly, the white metallic lustre of silver is absent in the black substance formed when it tarnishes. The tarnish, which is a corrosion compound, is silver sulfide, a compound of the elements silver and sulfur that forms by interaction between metallic silver, and hydrogen sulfide gas. Atomic theory explains this by stating that all matter is composed of very small units called '**atoms**'. All the units of one element or of one compound are identical to all the others making up that element or compound, but are distinct from those of any other element or compound. In compounds, an atom of each element is joined to other atoms in a special arrangement to form a **molecule** that is characteristic of the compound. The molecules of a single and pure compound are all identical – at least for simple compounds made from a very few elements. The links between the atoms in a molecule, which are called **molecular bonds**, will be discussed in the following chapters. The symbols for compounds are made up of the individual symbols of the elements of which each is composed (see Chapter 3).

C   Atoms and molecules

Molecules consist of two or more atoms bonded together. Usually the atoms are of several different elements: e.g. carbon dioxide contains carbon and oxygen atoms; acetone contains carbon, hydrogen and oxygen atoms (many other solvents do as well); silver tarnish contains silver and sulfur atoms. It is unusual for a molecule to contain more than five elements; two, three or four is the common range. Elements that are gases like oxygen, nitrogen and chlorine are mostly not found as single atoms but as molecules containing two atoms of the same element, or on occasion three, in the case of ozone which has three oxygen atoms. The atoms of elements that are solids at room temperature, such as metals, form up into much larger groups that determine the properties like ductility and conductivity (which are the ability to be worked into useful shapes, and the ability to carry an electrical current) that define a metal. In that sense one atom of a metallic element doesn't represent something that behaves like a metal, but huge numbers of them together do.

Although there is only a limited number of different atoms, and fewer than 40 elements which are common in compounds, the number of possible combinations even from only two or three of them is very great, especially when you consider that large numbers of several different atoms can be linked together in a specific order to form a molecule. We shall see later that the properties of this vast range of compounds are as much determined by the nature of the bonding between atoms as by the number and type of atoms themselves.

D   Solids, liquids and gases

The fact that substances can exist as **solids**, **liquids** or **gases**, and can change from one of these states to another, is also explained by the existence of atoms. For example, water is a compound which is liquid at ordinary temperatures, freezes at 0°C to become a solid, and boils at 100°C to become a gas. If you heat ice above 0°C it turns back to liquid and, similarly, by cooling the gas (steam) to below 100°C that too will turn back to a liquid, a fact used in purifying water by distillation. (A scientist might add, accurately though it can sound rather pedantic, that this is the case at sea level, meaning at the air pressure found on our planet at sea level. And then a mountain climber might recall that boiling water at altitude makes terrible tea. That is because it boils there at a temperature considerably below 100°C.) Rather a lot of statements in this introductory book should in fact conclude with a phrase like 'in a typical conservation workspace' to make them accurate.

The theoretical description of this is that molecules are firmly and closely stuck together in a solid, only loosely held together in liquids, and quite free to move independently from one another in gases. Already you will realise that some important properties are being explained. The rigidity of solids is accounted for by forces holding the atoms together. The observation that many solids occur as regular crystals with a repeating form is consistent with the idea of large numbers of a small range of elements stacked in a repetitive fashion. Regular stacking of many uniform shapes, like cans or chocolate boxes in a supermarket display, results in pyramids with edges, faces and points like those found in crystals.

The reason that liquids can flow is because these **inter-molecular forces** (that is, the forces between separate molecules) are less strong. That is why you can stir or pour liquids. Divers plunge into water with the confident expectation that the water molecules will move out of their way, while in winter conditions of 0°C or below the same water becomes capable of carrying the weight of animals and people because the water molecules are held rigidly together as ice.

In gases, mobility is even easier because the molecules are completely separated. A kilogram of water about to boil occupies just over one litre of space, pretty much the same space it occupied at room temperature. As steam at 100°C the same molecules in that kilogram of water steam can separate to reach all parts of the room.

Although atomic theory can explain the existence of the three physical states it has given no intimation of the role of heat in the transition from one condition to another. The hotter things are, the faster the particles move, and the more readily they can move away from one another. It takes an input of energy as heat, to transform a solid into a liquid and a liquid into a gas.

In gases, atoms and molecules move about rapidly, colliding frequently, and there is a great deal of empty space between them. They move more slowly in liquids and more slowly still in solids where they are packed tightly together, with just a little freedom to vibrate backwards and forwards. If the solid is made hotter, its molecules will move more rapidly. The faster they move, the more space they need to move in and so the substance expands and eventually passes from solid to liquid, and then, from liquid to vapour. The temperature at which the transition between solid and liquid occurs is known the **melting** or **freezing point** if the transition goes from solid to liquid, or liquid to solid respectively, and the temperature where the transition takes place between liquid to vapour is called the **boiling point**.

The observation that the smell of a solvent in an opened container can spread rapidly throughout the room from someone's table or trolley, is accounted for by molecules in the gaseous state, which are able to travel rapidly in the atmosphere in the form of vapour, even when most of the solvent is liquid. The closer the boiling point is to room temperature, the more molecules will be in the gaseous state, and the more **volatile** the solvent is. That in turn affects the risks it may pose to the conservator's health: a volatile solvent is more concerning in this respect, since its concentration in the workspace can become high, rapidly, and will remain high if the container is kept open.

As a demonstration of this fill a beaker with water. Then take a drop of a deeply coloured food colourant, or a brushload of watercolour paint, and drop it into the water. Do not stir the water. Even without stirring, the liquid gradually changes colour throughout. The simplest explanation of this is that the particles are in **random motion** and that it is this random movement of particles which has produced an evenly mixed solution in a liquid. In gases, the random motion is a lot faster, which is obvious when one is preparing onions to cook: their powerful smell is released as a gas. There is an obvious inference to make from these statements: the random motion gets faster as the temperature is increased.

E   Mixtures and purity

It has already been noted that there is a remarkable difference between a mixture of two separate elements and a compound made of the same two elements. In a mixture the elements would retain their individual properties rather than assuming the properties of the compound they might form. In a compound there are strong bonds between the constituent atoms. In a mixture, however, there are no chemical bonds between the components.

Generally speaking, all materials whether synthetic or naturally occurring are combinations of several kinds of molecules or atoms but, in mixing, these different particles have not undergone any chemical interaction or change. Different combinations exist at many levels of intimacy. An early twentieth-century house is quite often a 'mixture' of bricks, mortar, plaster, wood, nails, etc., though it seems perverse to regard it this way because the different materials are quite distinct, and each of these materials is there to serve a purpose. It only resembles a mixture if it has just been bulldozed to the ground. Lime mortar, which is a mixture of lime (calcium oxide and/or calcium hydroxide) and sand in water, is a more realistic example of a coarse mixture, because the separate components are purposely mixed together to make it, and it turns into a new compound distinct from its dry ingredients as it sets, or dries out. It is not an **intimate mixture** as is the result of some chemical reactions, since the sand is recognisable by close examination as lumps in the dried product. Concrete (Figure 2.1) is pretty obviously a physical mixture, which is made from aggregate and cement, and even when it is polished the lumps of aggregate are visible. On a finer scale, where the separate ingredients measure 1 micrometre and larger, up to the few centimetres' diameter of the aggregate in concrete, many of the materials that make up objects are not intimate mixtures either (Figure 2.2). As a rule of thumb, only things that look like a single material under very high magnification (quite a few hundred times) are at all likely to be intimate mixtures.

An understanding of these rather obvious, coarse mixtures is important to conservators. Often, in the course of your work, you will be confronted by an object composed of a variety of materials and you will be prevented from using a particular technique because the beneficial effects on one part of it will be outweighed by adverse effects on another. A particular solvent, for example, might have the unwanted effect of softening and dissolving a coloured layer as well as removing a discoloured coating from a piece of furniture now in a museum.

Figure 2.1 Concrete wall surface, Westminster underground station, London. This is a very coarse mixture, nothing like an intimate mixture. Image: Joyce Townsend

Figure 2.2 Multiple layers of mid-twentieth-century paint, viewed in cross-section. The orange and the pink paint layers, though well mixed, hardly qualify as intimate mixtures even though the coloured particles are small: it is obvious at this magnification that there are white particles as well as coloured ones in these layers. The larger black particles in the orange paint, separately added by mixing with a brush, certainly do not form an intimate mixture with the orange paint. Image: Joyce Townsend

It is worth noting that much of your work, all treatments described as 'cleaning' for instance, involves separating mixtures that have a junction between two or more substances – the object and the dirt, or the dirt on a coating on the object – and you will know very well that the difficulty of these jobs depends, to some extent, on how intimately mixed the two parts are. And that depends on whether they reacted chemically, or if they are only close together in the sense of being physically close together, or if they interacted physically so that there is no clear-cut boundary between them (an old shellac coating sinking into a plaster cast as it was brushed on, for example).

The most intimate mixtures are those at the level of atoms or molecules. Gases mix easily at the molecular level. The atmosphere is a good example of an intimate mixture of several kinds of atoms and molecules. Since air is more than 78% nitrogen you might call it impure nitrogen but that would miss most of its important properties. As a source of oxygen (oxygen forms another 21% of the air) and water vapour for human beings and carbon dioxide for plants, air is essential to life, but you would not face so many conservation problems if the sulfur compounds were absent. (Hydrogen sulfide is the agent which tarnishes silver, and sulfur dioxide causes the degradation of limestone masonry and other materials such as leather book bindings. Gases like these, and carbon dioxide, are typically present at parts per million or less, that is, 0.0001% or less for each gas).

Solutions form another common class of molecularly intimate mixtures. There are many familiar examples of substances dissolved in water – sugar in tea, salt in the sea, dilute acids and alkalis used to adjust the pH of some formulations for aqueous cleaning. Among the mixtures you make yourself during your work are those achieved by adding wetting agents (detergents, etc.) to water. If they dissolve and mix completely, it's a solution. Mixing two-component epoxy adhesives like the Araldite® series is an example (here the ingredients of resin and hardener are chemically interactive, but until they have set hard, they form a mixture, not a solution) and another example is adding pigments to adhesives and fillers (again, this is not called a solution, but it is a non-intimate mixture).

Solid solutions are molecularly intimate mixtures that were made at high temperatures, but which are solid like minerals or metals when they have cooled to room temperature. Alloys of gold with silver added would be good examples.

It is rare to find a substance that contains only one kind of atom or molecule. The most likely place you might expect to find matter containing only one kind of molecule is among the jars and bottles of chemicals in your conservation studio or laboratory (though not the commercial products). Most of these will be examples of compounds. If it were possible to have a specimen with only one kind of molecule present it would be referred to as a pure compound. There will, however, invariably be some molecules of other compounds present, and these are called impurities. Very many reagents and solvents can be purchased at a **purity** of 99–99.5%, as 'general reagents' or 'industrial grade', which are not regarded as being sufficiently pure for many scientific uses, and these have a (relatively) low price. More pure and much more expensive versions of the same material would be labelled 'Analar' or 'Ultrar' or 'spectroscopic grade'. This is a quality rating; the label will inform you what percentages of which of the major impurities are present. A price list of chemicals from one of the suppliers will usually list the following categories:

- Industrial grade is usually quite adequate for conservation purposes such as surface cleaning, unless it is known that an unwanted reaction with one of the impurities can occur. With organic solvents such as acetone there is very occasionally an undesirable residue after evaporation.
- AnalaR grade is usually old stock and better than 99.9% purity, in fact better than needed for most conservation treatments.
- Spectrographic grade is only needed for sample preparation for chemistry-based materials analysis where it is important that no stray compounds interfere with the detection and measurement of very small quantities of material. Most conservation treatments do not demand such purity in the reagents and chemicals used.

Generally, because of the effort involved in removing the last traces of impurities, you will pay a lot more for a very nearly pure chemical than one where there is a higher proportion of impurities. Compare the price of supermarket washing soda with that of spectrographic grade sodium carbonate.

F   Physical and chemical changes

Changes in the condition of materials are always important to the conservator. The deterioration of objects to a state where they need active conservation treatment is the result of change. You work to put obstacles in the way of those changes if that can be done by protecting surfaces with a coating for example, to make them occur at a slower rate in future, and sometimes to reverse them for a time. The changes during both deterioration and conservation treatment can be classified as either **physical changes** or **chemical changes.** A physical change of condition involves some alteration at a microscopic level, without any change in the structure of the individual molecules. Chemical changes involve rearrangements of atoms among molecules to create new molecular structures. The components of a mixture may be separated by a physical change (boiling and collecting them separately for example, which is distillation), whereas when the atoms in a molecule are permanently separated and rearranged in a new way, a chemical change has taken place.

A great deal of conservation treatment uses physical changes. If you blow or gently brush or vacuum clean the dust off a museum exhibit you have merely moved the dust from one place to another, you haven't wrought any chemical changes upon its molecules. Similarly, when a wax/resin mixture is used as a thermoplastic adhesive (for example, in attaching a textile-based strip lining to the weakened canvas turnovers of a painting on canvas, or in the past by lining another canvas to the entire canvas support) it is melted by heat and flows between into the two fabrics. When it cools the molecules of the adhesive mixture cease to move and begin to hold the fabrics together. This is a physical change, but one that cannot always be reversed simply by heating the two canvases to the same temperature used for the lining process: if some of the adhesive moved into the paint film – or did so later, influenced by the surrounding environment – it might influence the chemical changes that occur as the paint ages. Other steps like repeated heating and blotting might be needed to remove most or all of the adhesive. Another physical change occurs when a solvent-borne coating, such a solution of Paraloid® B-72 in acetone, dries on a surface. The solvent, acetone, evaporates, leaving a continuous film of resin behind. The solvent molecules have left the surface and the resin molecules have remained; there has been no rearrangement within the molecules.

The tarnishing of silver, however, is a chemical change. Silver atoms combine with sulfur atoms to form black silver sulfide. In typical environmental conditions in museums, the reaction always happens this way, and never the other way: tarnish does not revert itself to metal. The corrosion of bronze and the rusting of iron are also chemical changes that can take place in both museum and outdoor conditions. In these two examples, the metal is simply turning back into the metal ore it was made from in the first place, one that had lasted not just for thousands of years, but over geological time: that is the stable form, but the metal is less stable.

It is not always easy to differentiate between physical and chemical changes. Cleaning processes involving solvents to remove a coating from an object are examples where the distinction between physical and chemical changes is more blurred. Both may be involved, and a physical change such as swelling of the material to be removed by a swab stick before it turns back into its original film may be brought about by a chemical reaction (solubility of the coating in the solvent). (Consideration of these processes has been left for the book in this series titled *Cleaning*).

G   How chemical reactions happen

During the course of this chapter you will have come to think of the molecules of a compound as atoms bonded together into characteristic patterns. Chemical change has been explained as a transfer of some individual atoms among molecules. To understand how these changes can happen you need to know what causes the atoms to break their bonds to form new patterns.

When a substance is heated the atoms and molecules all move with increasing speed, and so, as you can imagine, they run increasing risks of colliding, and with greater impact each time they do collide. As they collide with one another the molecules may join together or knock single atoms or fragments off each other. These fragments may recombine to form new combinations. The more stable (strongly formed and strongly bonded) a molecule is, the stronger the forces that are needed to break it up and cause it to react chemically, that is to make new combinations. This concept of molecules in constant movement, with the movement increasing with the increase of temperature, explains why chemical reactions go faster at higher temperatures. As the collision speeds get higher, molecules may become capable of breaking at more points in their structure, creating a greater variety of fragments that are then free to combine in more and different ways, and to form a larger range of fresh molecules. It is also likely that while parts of molecules are randomly moving and colliding together they may temporarily bond to form an unstable mass which easily falls apart to form yet other molecules which may be more stable (firmly bonded together). That level of chemical detail is usually worked out much later than the understanding that the stable product is the 'result' of heating the ingredients together.

As a conservator, you will quickly understand from this that there are good reasons why, if the instructions for a chemical treatment of an object specify a temperature, you should keep to it. Otherwise the proposed reaction may go too fast to allow control, or reactions different from those intended may occur. If the

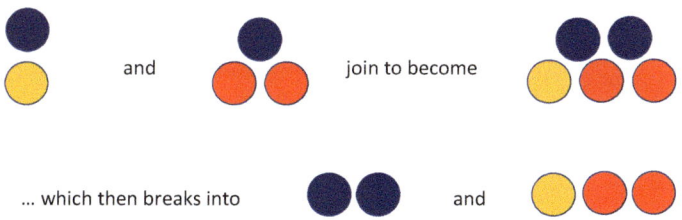

Figure 2.3 One possible way in which three sorts of atoms may collide, join and then break up during a reaction, to form different molecules.

treatment were carried out in the field in a hot climate, or on an archaeological site in winter, these temperatures would make it go at a different speed from that in a comfortably controlled workspace. At higher temperatures in some climates, processes such as brushing on a coating consisting of a polymer applied in a solvent might be less successful, because the solvent might evaporate before the coating could level out and become free of brushmarks. Even spraying a varnish with a spray-gun on a very hot day can result in a rather opaque-looking, non-smooth coating because too much solvent evaporates before the droplets of resin in solvent can form a smooth coating.

3   Molecules and chemical equations

Introduction

Different ways of thinking about and drawing molecules are introduced here. These help with understanding how molecules interact, and how that reaction can be described by a chemical equation that also aids understanding of the whole process. Examples are given, and exercises test the reader's understanding.

A   Visualising molecules

Most people find it easier to understand something if they can hold some sort of visual image in their minds. Individual molecules are much too small to be seen even with the most powerful optical microscopes. The smallest spot you can see through a good optical microscope is about ten thousand atoms in diameter. Atomic force microscopy, which can give a very good impression of how atoms are arranged in a solid, or X-ray diffraction carried out using synchrotron radiation (which can in effect generate an image at a much finer scale than the X-ray diffraction instruments used in conservation science laboratories), both provide a visualisation of atomic arrangement, rather than an image of an isolated, individual atom. It is perhaps the wrong question to ask what molecules 'look like', just as it is not sensible to ask what colour they might have. Nevertheless, they do have a three-dimensional reality which must be related to what we call the 'shape' or 'arrangement' of visible objects, and to their properties as well.

The individual atoms of which molecules are composed can be thought of as spheres, because that is the simplest shape and it has equal properties in all directions. However, in molecules the distance between the centres of these spheres is influenced by the neighbouring atoms: the spheres get pulled out of their symmetrical shape. A realistic way to show the shape of a molecule is shown in Figure 3.1.

Known facts about the behaviour of gases are consistent with this **molecular model**, and suggest that they do have shape and volume, but because they are capable of being modified by the presence of their neighbours, they do not have rigid outlines.

Methane, the main component of natural gas, has molecules that contain four hydrogen atoms and one carbon atom. It is known that none of the four hydrogen atoms is joined to another hydrogen atom but that each is bonded individually to a

DOI: 10.4324/9781003261865-3

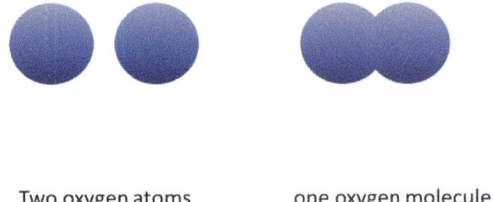

Two oxygen atoms   one oxygen molecule

Figure 3.1 Schematic representation of (left) two separate oxygen atoms and (right) one oxygen molecule where the two oxygen atoms are connected.

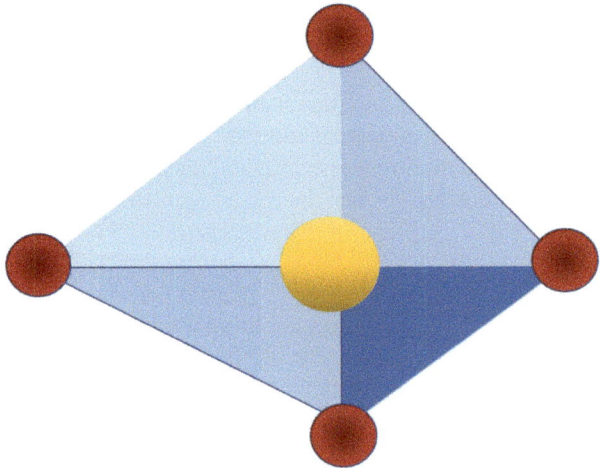

Figure 3.2 Schematic representation of one methane molecule, looking like a spiky sphere.

central carbon atom. The hydrogen atoms surround the carbon atom, keeping as far away from each other as possible. This means that the hydrogen atoms are at the corners of a pyramid with a triangular base (a tetrahedron) with the carbon atom in the middle. It is the closest that these five molecules can get to forming a sphere together. Figure 3.2 immediately shows the difficulty of representing this knowledge in 2 dimensions even for so simple a molecule. The most realistic way to portray three-dimensional entities is to use three-dimensional models. These are often used for teaching science and are of two types.

'Ball and stick' models show the links (bonds) between atoms clearly, and give some sense of three dimensions, as in Figure 3.3(a), but obscure the fact that in reality the atoms merge together in some sense, as the space-filling model in Figure 3.3(b) shows. Although both types of model provide useful information they may also suggest things that are probably not true of the actual molecules. For instance, in both types of model the atoms of different elements are shown in different colours and there is an obvious division between the different atoms. The models are sometimes designed to come apart to show how molecules can decompose and

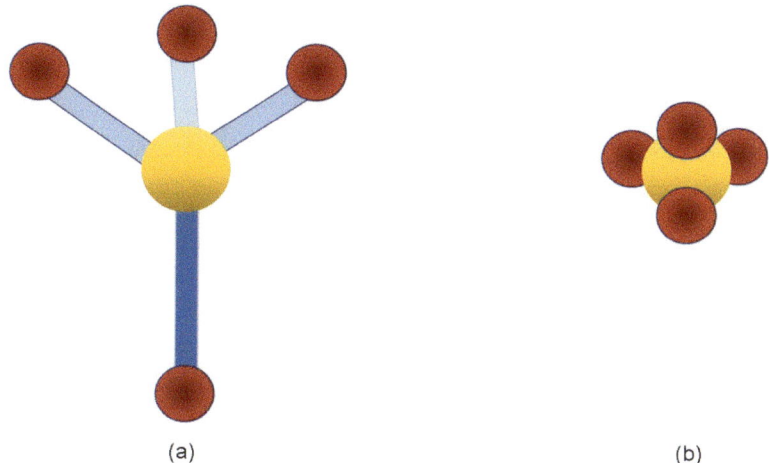

(a)   (b)

Figure 3.3 Three-dimensional models of methane molecules: (a) the 'ball-and-stick' model
and (b) the 'space-filling' model.

rearrange, but the links are always rigid, which obscures the fact that bonds in
molecules in fact vibrate by stretching out and back, and in fact are never static at
everyday temperatures.

B   Symbolic representations of molecules

Owing to the practical difficulties of depicting molecules accurately, more symbolic
ways of representing them are used by scientists. These are much more useful than
the ones we have just seen for discussing chemical reactions. Consequently, they
are the ones most frequently found in chemistry books and in conservation texts.

B1   *Molecular formulae*

The first of these more symbolic models is the **molecular formula**. ('Formula'
takes the Latin plural 'formulae'.) It indicates concisely how many atoms of exactly which elements are contained in each molecule of a compound.

   The molecular formula for methane is CH_4. Comparing this combination of two
letters and a number with the pictures in Figures 3.2 and 3.3 shows you immediately that it describes the molecule as having one carbon atom and four hydrogen
atoms. The elements present are identified by their symbols (as discussed in Chapter 2), and there is another convention to show how many of each sort of atom is
present. Although the information might have been written CHHHH, the convention is to represent any number of atoms greater than one by a number written to the
bottom right of the element's symbol, as a **subscript**. Thus in CH_4 the C stands for
one atom of carbon and H_4 stands for four atoms of hydrogen. (The symbols must
always be written carefully and with due regard to capital letters; Co is the symbol
for the element cobalt, but CO is the molecular formula for carbon monoxide.)

Exercises

In order to help you become familiar with writing molecular formulae it may be useful for you to do the following:

1 Write down the appropriate numbers in the blank spaces in the following examples:

   (a) The molecular formula for carbon dioxide is CO_2. Each molecule contains_____atom(s) of carbon and_____atom(s) of oxygen.
   (b) The molecular formula for sulfuric acid is H_2SO_4. Each molecule contains_____atom(s) of hydrogen, _____atom(s) of sulfur and_____atom(s) of oxygen.
   (c) The molecular formula of acetone is C_3H_6O. Each molecule contains_____atom(s) of carbon, _____atom(s) of hydrogen and_____atom(s) of oxygen.

2 Write down molecular formulae for the following molecules:

   (a) Water, which contains two atoms of hydrogen and one atom of oxygen.
   (b) Ammonia, which contains one atom of nitrogen and three atoms of hydrogen.

B2   *Structural formulae*

Molecular formulae only give limited information, however, and do not, for instance, tell us which atoms are joined to which. This information is conveyed in another representation – the **structural formula** (the plural is 'structural formulae'). Figure 3.4 shows the structural formula for methane, where each atom is represented by its symbol and the bonds between them are shown as dashes, spaced apart equally. Although this shows which atom is joined to which, it is still a flat and static representation of what is in reality a three-dimensional entity, but it is easier to draw.

Consequently, you will sometimes see attempts to present a more three-dimensional view within structural formulae, such as Figure 3.5, where again the four hydrogen atoms and carbon atom are shown as equally spaced apart.

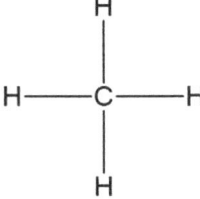

Figure 3.4 Structural formula for methane.

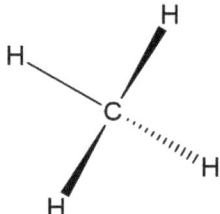

Figure 3.5 Another representation of methane, which uses a form of perspective to suggest the molecule's three-dimensional shape.

With larger, more complicated molecules, their shape actually dictates how they react with other molecules and so a two-dimensional representation of a three-dimensional shape is often less helpful. Biochemistry and biological processes often involve very large molecules whose shapes determine how and whether they can react together.

B3   *Structural formulae and isomers*

The molecular formula for ethanol (the major constituent of industrial methylated spirits (IMS), is C_2H_6O. Figure 3.6(a) shows its structural formula, while Figure 3.6(b) is also a representation of C_2H_6O composed of the same numbers of carbon, hydrogen and oxygen atoms, but it is not the structural formula for ethanol. When the atoms are linked together as in Figure 3.6(b) the compound is called dimethyl ether. Dimethyl ether was once used in conservation treatments, but its toxicity would prevent this today: here it is used only to illustrate a concept.

There are many groups of compounds that have the same molecular formula but have different structures. Such compounds are called **isomers**.

There are three possible isomers of C_2H_4O. One is called ethylene oxide, shown in Figure 3.7(a) and another is acetaldehyde, shown in Figure 3.7(b).

Ethylene oxide is a gas which used to be legally permitted in European countries to kill insects infesting objects or packing materials in museums, and which may

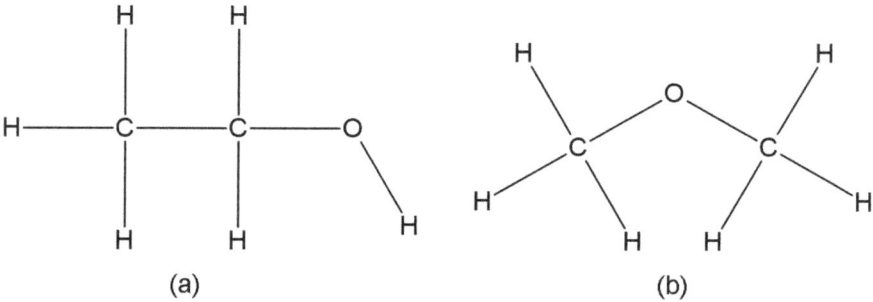

(a)   (b)

Figure 3.6 Two structural formulae for C_2H_6O.

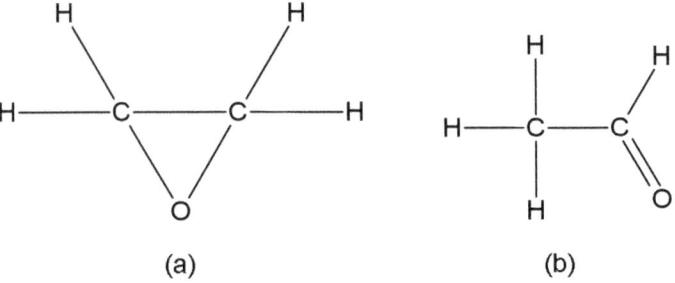

Figure 3.7 Structures for C_2H_4O.

be still legal for such use in some countries. When it is heated, ethylene oxide is converted into its isomer acetaldehyde.

Two new features are shown in Figure 3.7. In ethylene oxide the carbon and oxygen atoms are joined together in a way that is more ring-like than linear, which is quite a common structure. In the formula for acetaldehyde there is a double dash between the C and the O. This means that the bond between C and O in this molecule is not the same as the C–O bonds in ethyl alcohol, ethylene oxide or dimethyl ether, but is called a **double bond**. The different kinds of links between atoms will be explained more fully in the next chapter. The point which it is important for you to understand now is that two or more molecules which have the same molecular formula can show quite different chemical properties because they have different structures. With experience you will come to anticipate some of the properties of a particular molecule by looking at its structural formula.

The fact that chemical behaviour depends on structure as well as which atoms a molecule contains, leads to a further development of molecular formulae to indicate more about molecular structure, by grouping together the smaller units. The molecular formula for ethanol can be written as C_2H_5OH, that of dimethyl ether as CH_3OCH_3 and that of acetaldehyde as CH_3CHO. These compact representations give almost as much information as the full structural diagrams. Unfortunately ring compounds such as ethylene oxide cannot be easily represented by a string of symbols: the shorthand works better for a chain of small units.

C   Building chemical equations

In the preceding section you have seen representations of oxygen and methane molecules. Both are gases at room temperature, oxygen being a constituent of the air, and methane being a major component of natural gas. When a gas burner is lit a chemical reaction is started between these two gases. Beyond the edge of the flame you would be unlikely to find any methane molecules. Instead, there are two different gases, water (found as a gas because of the heat of the flame) and carbon dioxide. The **chemical reaction** between them is the process by which methane and oxygen become carbon dioxide and water.

This could be written down in words:

Methane mixed with oxygen and ignited turns into carbon dioxide mixed with water.

Obviously such descriptions rapidly become cumbersome while imparting only limited information, and so symbols are used instead. 'Mixed with' can be denoted simply by +, 'turns into' by →. Special instructions like 'ignited', in this case a description of the necessary temperature to affect a reaction (say 500°C), are written on the arrow.

Methane + oxygen $\xrightarrow{500°C}$ carbon dioxide + water

The next step is to use molecular formulae for the substances. Carbon dioxide is CO_2 (one atom of carbon and two of oxygen), water is H_2O (two atoms of hydrogen with one of oxygen) and so what might be written is this:

$CH_4 + O_2 \xrightarrow{500°C} CO_2 + 2H_2O$

In this form, the description of the reaction starts to become useful for understanding more about what is happening. For example, if the chemicals were weighed before and after the reaction, assuming no part of any substance had been lost, you would find no change in weight. The weight of the **reactants** (the chemicals you start with) is equal to the weight of the **products** (what you finish with). This is because no atoms are destroyed or made during the reaction; the event is just a rearrangement of the atoms you first had. The **chemical equation** (as this symbolic description of a reaction is called) should show that the number of oxygen atoms is unchanged and so is the number of hydrogen atoms, although all the atoms are combined differently with one another. Look at it again:

$CH_4 + O_2 \xrightarrow{500°C} CO_2 + H_2O$
methane + oxygen $\xrightarrow{500°C}$ carbon dioxide + water

Think of the formulae not as mere shorthand for the names, but as representing one *molecule* of each chemical. You will then see that the four hydrogen atoms at the start appear to have turned into just two and that the two oxygen atoms have increased to three at the right-hand side of the equation. In order to balance the chemical equation and maintain the same number of atoms on each side, it is necessary, therefore, to find the number of molecules which react together to change methane and oxygen into carbon dioxide and water without any implied destruction or creation of atoms. This will work:

$CH_4 + 2O_2 \xrightarrow{500°C} CO_2 + 2H_2O$
one molecule of methane + two molecules of oxygen →
one molecule of carbon dioxide + two molecules of water.

Now count the atoms of each kind on each side of the reaction:

Table 3.1 Atoms involved in the burning of methane in oxygen.

Element	Atoms present	
	Before	*After*
C	1	1
H	4	$2 \times 2 = 4$
O	$2 \times 2 = 4$	$2 + (2 \times 1) = 4$

This complete description tells you how many molecules of different sorts react together. It is called a **balanced chemical equation** when the number of each type of atom is the same on either side of the arrow. Balanced equations used to be shown with an 'equals' sign as $=$ instead of the arrow which is why we have the word 'equation', that has the same origin as the phrase 'equates to'. Whenever a chemical reaction is represented by a chemical equation it should be balanced, because otherwise it does not convey all the information that it could convey. The arrow conveys the direction of the reaction in certain circumstances, such as heating the materials to 500°C, so it gives more information. (If no temperature is mentioned, chemists have always assumed that a rather warm room temperature of 25°C is meant).

All this assumes there was as much oxygen available as needed. If it is in limited supply, the reaction might not proceed this way: carbon monoxide might be formed instead.

As the next section will reveal, balanced equations allow an interpretation of events on the minute scale of atoms and molecules. The ability to understand these descriptions of chemical reactions will often be of great value to you in your work.

Exercise

3   Try balancing the equations for the reactions in which propane (C_3H_8) and butane (C_4H_{10}) are burned in oxygen to form carbon dioxide and water.

D   Chemical equations in use

Chemistry can help you understand more thoroughly the nature of the changes which occur when an object is manufactured, as it ages and deteriorates, and when it is treated. It can suggest why objects made of particular materials are subject to particular forms of change; you can use this knowledge to reduce or prevent further damage and to judge the suitability of alternative conservation treatments.

This section gives two examples to show how your knowledge gained through observation can be illuminated further by knowing some chemistry.

D1 The manufacture and deterioration of wall paintings a fresco (true fresco)

Listed below are the chemical names and molecular formulae for all the materials which are introduced in this section. Although you may find these a little foreign to you at present, much more will be said about how these formulae are arrived at and what the names mean in Chapters 4 and 5 respectively. This need not worry you at present, although it may help you to refer to Table 2.1 (p. 23) so that you can remind yourself of the shorthand symbols for the various elements listed. The last four substances in the list are solids at room temperature. You will learn later that it is more accurate to think of them as having an extended structure, rather than consisting of individual molecules; but their molecular formulae can be used in chemical equations in exactly the same way as simple molecules like methane and oxygen.

In the fourteenth-century *Il Libro dell'Arte/The Craftsman's Handbook* by Cennino D'Andrea Cennini, in the section headed 'The Method and System for Working on a Wall, that is, in Fresco', the preparation of the wall (application of the first plastering, the *arricio*) is described as follows:

> When you want to work on a wall, which is the most agreeable and impressive kind of work, first of all get some lime and some sand, each of them well sifted … and wet them up well with water, and wet up enough to last you two or three weeks. And let it stand for a day or so, until the heat goes out of it: for when it is so hot, the plaster which you put on cracks afterward. When you are ready to plaster, first sweep the wall well, and wet it down thoroughly, for you cannot get it too wet. And take your lime mortar, well worked over, a trowelful at a time; and plaster once or twice, to begin with, to get this plaster flat on the wall. Then, when you want to work, remember first to make this plaster quite uneven and fairly rough.

Table 3.2 Reactants involved in the making of true fresco and its later deterioration in polluted cities. The symbol (OH)$_2$ indicates that there are two of the OH group.

Chemical name	Common name	Molecular formula
water	water	H_2O
sulfur dioxide	sulfur dioxide	SO_2
sulfur trioxide	sulfur trioxide	SO_3
sulfuric acid	sulfuric acid	H_2SO_4
carbon dioxide	carbon dioxide	CO_2
silicon dioxide	sand, silica	SiO_2
calcium oxide	lime, quicklime	CaO
calcium carbonate	chalk; limestone; calcite; (these are different materials)	$CaCO_3$
calcium sulfate	calcium sulfate; (when used to make a plaster mould or support) plaster of Paris	$CaSO_4$
calcium hydroxide	slaked lime	$Ca(OH)_2$

When the *arricio* had dried (after some days) a second layer of plaster (*intonaco*) was applied on the morning of the day on which the painting was to be done; the area that was covered was limited to the amount that the artist knew that he could complete in a day's work, and it might have been a single component or one figure in a larger composition. The artist then painted pigments mixed with water over the wet *intonaco*.

If you look at the list given at the beginning of this section, the raw materials for the plaster (lime, sand and water) are the chemicals calcium oxide, silica and water; and yet when a piece of historic true fresco is analysed chemically the only materials found are silica and calcium carbonate. Clearly some rearrangements of atoms among molecules (that is, chemical reactions) have taken place.

Here are the equations describing the chemical reactions which occur:

$$CaO + H_2O \rightarrow Ca(OH)_2$$
$$Ca(OH)_2 + CO_2 \rightarrow CaCO_3 + H_2O$$

Before you read on, look again at Cennini's account, and at the list of molecular formulae. Using the information in these and your knowledge of chemical reactions see whether you can:

- Say what these two reactions are in words.
- Deduce when the first reaction takes place.
- Say where the carbon dioxide involved in the second reaction might come from.
- Decide whether sand is involved in the reactions.
- Say where the water has gone.

The first reaction to occur takes place when the sand (SiO_2) and lime (CaO) are mixed with water, and slaked lime ($Ca(OH)_2$) is formed. More water than is necessary for the reaction is used so that the plaster is a workable paste that can be spread onto a wall. Looking at the equations above, the formula SiO_2 does not appear. This implies that the sand constituent of plaster does not take part in either of the chemical reactions; it is only there to help strengthen the plaster, and at the end it will still be possible to recognise it as sand. The second reaction describes one 'molecule' of calcium hydroxide reacting with one molecule of carbon dioxide and producing one 'molecule' of calcium carbonate and one of water. The carbon dioxide involved in this reaction comes from the air and the reaction takes place over a period of months as air diffuses into the plaster. It may even take years for the hydroxide to be converted into carbonate through the whole thickness of the *arricio*. Spare water gradually evaporates from the wall as the plaster dries out. It is probably the spaces left by the water which provide porosity in the plaster through which air can penetrate to produce the second reaction.

There are many frescoes (especially, for example, in Venice) made in the way described, and which have suffered serious deterioration in the fairly polluted atmosphere of the city. The main pollutant during the twentieth century was sulfur dioxide generated by burning the small quantity of sulfur present naturally in fossil fuels (coal, oil and natural gas). In a polluted atmosphere two reactions take place. Refer to Table 3.1 to determine what the names of the reactants and products of the two reactions are:

$$SO_2 + O_2 \rightarrow SO_3$$
$$SO_3 + H_2O \rightarrow H_2SO_4$$

However, one of these equations is not balanced. Can you deduce which of the equations it is and how it can be made to balance?

The first equation has one more oxygen atom on the left-hand side than on the right. To make it balance two molecules of sulfur dioxide need to combine with one molecule of oxygen forming two molecules of sulfur trioxide as the product of the reaction.

Demonstration

Take a chip (about one cm^3) of marble, limestone or chalk, which are all calcium carbonate, and some dilute acetic (ethanoic) acid. (Handle acids with care, and whilst wearing eye protection and acid-resistant gloves. Plan in advance how the acid will be disposed of: in this case, by turning on the cold water, and gently pouring it down the sink while wearing the eye protection and gloves, and letting the cold water run for several minutes more.) Try putting the chip in the acid and leaving it there for a few hours. What evidence of chemical reaction is there to be seen?

The reaction between calcium carbonate and sulfuric acid is described by the equation:

$$CaCO_3 + H_2SO_4 \rightarrow CaSO_4 + H_2O + CO_2$$

Observing this demonstration, you will have seen bubbles of gas rising from the chip and eventually found that the chip has dissolved in the acid. The reaction is one way of helping to identify calcium carbonate, and hence marble or limestone in a rock formation outdoors. The demonstration also shows you in action how the decay of frescoes, marble statuary and lime mortar in brickwork can be accelerated in the presence of sulfuric acid. Clearly the best long-term preventive measure would be to take steps to reduce the emission of sulfur dioxide into the air, and the increased use of sustainable energy sources and decrease in the use of coal to generate electricity is achieving precisely that. Another option used in the past has been to remove the object to a museum that has air conditioning that used a filtered air intake. However, the problems that face the conservator are immediate. Knowing the chemistry of the problem tells you that the object must be kept dry and that polluted air should be excluded from it if further deterioration is to be prevented.

Table 3.3 Reactants involved in the darkening of underbound lead white pigment and its reversal by bleaching with hydrogen peroxide. This treatment, once very common in many conservation specialisms, may not be appropriate for lead white-based paint on certain substrates such as paper.

Chemical name	Common name	Molecular formula
water	water	H_2O
hydrogen peroxide	peroxide	H_2O_2
carbon dioxide	carbon dioxide	CO_2
hydrogen sulfide	hydrogen sulfide	H_2S
lead carbonate	lead carbonate	$PbCO_3$
basic lead carbonate	lead white	$Pb_2CO_3(OH)_2$
lead hydroxide	lead hydroxide	$Pb(OH)_2$
lead sulfide	lead sulfide	PbS
	(as the mineral) galena	
lead sulfate	lead sulfate	$PbSO_4$

D2   The blackening and subsequent treatment of underbound lead white pigment

All the compounds shown in Table 3.3 appear in this example and are listed here as a guide when reading the related text.

Lead white, as you can see from the list, is in reality a single compound ($Pb_2CO_3(OH)_2$), although it can be considered as reacting as if it were two distinct compounds ($Pb(OH)_2$ and $PbCO_3$. It is therefore sometimes written in terms of the two compounds, thus, $PbCO_3.Pb(OH)_2$, linked by a full stop but no space between them to indicate that they are both part of the same compound.

Lead white is not an uncommon pigment in wall paintings and decorative paintings produced by many cultures in different time periods, and such paints might by now be rather lean (also known as being underbound), with very little paint medium surviving at the surface after natural ageing and/or weathering. Or they might have been mixed lean from the start, so that later layers of paint would be absorbed slightly, and key in successfully. Unfortunately lead white may chemically react in an atmosphere polluted with hydrogen sulfide gas, and form black lead sulfide as a thin surface layer. Hydrogen sulfide has often been present in traces in the air, historically more so than today. It is released during the decomposition of plants and animals, and from their dung which might have been burnt as a fuel; it is emitted by active volcanoes, and it is a toxic gas with a characteristic smell of rotten eggs.

The (balanced) equation for the reaction between lead carbonate and hydrogen sulfide to produce lead sulfide is:

$$PbCO_3 + H_2S \rightarrow PbS + CO_2 + H_2O$$

The equation for the reaction between lead hydroxide and hydrogen sulfide is:

$$Pb(OH)_2 + H_2S \rightarrow PbS + 2H_2O$$

Adding these two descriptions together gives the equation for lead white reacting with hydrogen sulfide:

$$PbCO_3.Pb(OH)_2 + 2H_2S \rightarrow 2PbS + CO_2 + 3H_2O$$

In the recent past and sometimes today when the original colours have great significance (botanical illustrations for example, where the colours must accurately represent the real plant), works on paper disfigured by their lead white having turned black are treated with hydrogen peroxide (H_2O_2). This is a very unstable compound and its molecules easily fall apart releasing oxygen and water. It is effective in this case because the freed oxygen reacts with the molecules which produce the colour. Its action on lead sulfide is to convert this black matter to lead sulfate which is white too, although it is not the original pigment.

$$PbS + 4H_2O_2 \rightarrow PbSO_4 + 4H_2O$$

The two examples just discussed show how the conservator can be assisted by chemical insight into the causes of deterioration. The ability to understand chemical equations can give you precise and concise descriptions of what is occurring.

Exercises

4   Lime is prepared from limestone or chalk which is heated in a lime kiln. The considerable heat decomposes the calcium carbonate (of which both chalk and limestone are composed) into calcium oxide (lime) and carbon dioxide gas is given off. The kiln is usually heated by a coke burning oven. The lime sinks to the bottom of the kiln and is removed. Write down the chemical equation for this reaction.

5   Suggest a possible reaction between calcium carbonate and sulfuric acid. Calcium sulfate is one of the products. Write the chemical equation.

6   This reaction between calcium carbonate and sulfuric acid may be used to distinguish between calcium carbonate and calcium sulfate. There is no reaction between sulfuric acid and calcium sulfate. The reaction might be used to determine whether the material in the ground of a painting is chalk or gesso, that is, whether it is calcium carbonate or sulfate. A small sample of the material in question is placed on a microscope slide and a drop of dilute sulfuric acid ($0.1M H_2SO_4$) or dilute hydrochloric acid ($0.1M HCl$) is added. If the material being treated is calcium carbonate then bubbles can be seen rising from the sample floating in the acid. No such bubbles are visible if the sample is calcium sulfate. What are these bubbles?

E   Making chemistry quantitative

E1   *Atomic and molecular quantities*

A molecular formula tells you what a material is made of, but when this knowledge is applied to a conservation problem it is often necessary to know the quantities of the materials involved in the treatment. You may, for example, want to know how much of one substance will react with a certain amount of another. Realistic quantities of the materials you handle can be measured by weighing in the normal way. Balanced equations, on the other hand, refer to small yet very precise numbers of molecules. They describe the reactions of such minute quantities of substances that it is impossible to employ the usual practical procedures for measuring. To derive weights of materials from balanced equations requires the reasonable supposition, the basis of atomic theory, that atoms can be well defined and that all atoms of the same element weigh the same, that is, they have a particular **atomic mass**. Likewise, a specific molecule has a well-defined **molecular mass** which can be calculated from the mass of the atoms that it contains. You could not make measurements to calculate how much lime would be produced from a given amount of chalk using just one molecule of calcium carbonate, but the amount of lime produced would be a definite percentage of the amount of chalk from which it was made. The equation for a whole lump of chalk being converted to lime would be:

$$nCaCO_3 \xrightarrow{\text{heat}} nCaO \quad + \quad nCO_2$$

| n molecules of | $\longrightarrow$ | n molecules of | + | n molecules of |
| calcium carbonate | | calcium oxide | | carbon dioxide |

where n is used to represent an enormous number.

Before we can get any further with this problem you must learn something about the weights of individual atoms. Before we can do this, you must understand that there is a slight but important difference in meaning between the terms weight and mass. **Mass** is a measure of the amount of matter and is a universally constant property of matter. **Weight** is dependant on the local gravitational field, although to a good approximation in everyday life, this is the same all over our planet, making it have the same numerical value as mass. But a given mass will have a different (and lower) weight on the moon. Although the words mass and weight are often used interchangeably in conservation, it is more correct to refer to **atomic mass** and **molecular mass** rather than atomic weight and molecular weight.

In the early days of chemistry it was a problem to make consistent measurements of the masses of atoms and molecules. Now there is a very accurate scale of the masses of the different elements' atoms relative to one another. This is known as the scale of relative atomic masses although it is often (and less precisely) referred to as the scale of atomic weights. For your practical purposes, however, the minute accuracy available from modern techniques is not often needed, and a simple scale of **relative atomic masses** will be used in this book. (It ignores the existence of **isotopes**: that is, different and minority 'versions' of an element that are heavier because they have a greater number of neutrons than the majority version, but the same number of protons – which are terms that will be discussed later, in Chapter 4. The consequence of isotopes existing is that the accurate atomic mass of each element is not a round number as we shall assume hereafter, but a slightly larger number than the round number, though still a well-defined value for each type of atom.) On this scale of relative atomic masses, a hydrogen atom, the lightest of all atoms, has a mass of one unit. The atomic masses of the other elements are expressed as multiples of this. Table 3.4 shows a list of atomic masses for the more common elements. The figures are given to the nearest whole number, as the margin of error in doing this and ignoring the existence of isotopes is usually less than one percent.

With this set of atomic masses and the knowledge you have already acquired of molecular formulae, you will be able to work out how heavy one molecule is compared with another. These **relative molecular masses** (familiarly, but not so correctly, called molecular weights) are found by adding together the atomic masses of all the atoms contained in each molecule, using the molecular formula to tell you how many times the atomic mass of each element has to be added in. Going back to the lime-burning equation, for example:

$$CaCO_3 \xrightarrow{\text{heat}} CaO + CO_2$$

the molecular mass of calcium carbonate ($CaCO_3$), can be calculated by adding together the atomic masses of one atom of calcium, one atom of carbon and three atoms of oxygen (because there are three oxygen atoms in the formula). The total is the molecular mass of calcium carbonate.

Thus:

$1 \times Ca$	atomic mass		= 40
$1 \times C$	'	'	= 12
$3 \times O$	'	masses at 16 each	= 48
Molecular mass of CaCO3			= 100.

Table 3.4 Approximate atomic masses of the more common elements.

Element	Chemical symbol	Atomic mass	Element	Chemical symbol	Atomic mass
aluminium	Al	27	lead	Pb	208
antimony	Sb	122	nitrogen	N	14
calcium	Ca	112	oxygen	O	16
carbon	C	12	potassium	K	39
chlorine	Cl	35	silver	Ag	108
copper	Cu	64	sodium	Na	23
gold	Au	197	sulfur	S	32
hydrogen	H	1	titanium	Ti	48
iron	Fe	56	zinc	Zn	65

Similarly, for quicklime (CaO) and carbon dioxide (CO_2) the simple sums can be laid out:

1 × atomic mass of Ca at 40	= 40
plus 1 × atomic mass of O at 16	= 16
Hence, the molecular mass of CaO	= 56.

And

1 × atomic mass of C at 12	= 12
plus 2 × atomic mass of O at 16 each	= 32
Hence, the molecular mass of CO_2	= 44.

You can now see the logic of the argument. The quantity of lime made from limestone is 56% by weight (or, **weight percent**) because each 'molecule' of calcium oxide weighs 56/100 of the calcium carbonate 'molecule' from which it comes.

Exercise

7   Using the same process and consulting Table 3.3 work out molecular masses for

(a) methane,
(b) oxygen molecules and
(c) water.

E2   *Molar quantities*

Look again at the balanced equation for the combustion of methane shown earlier, and then at the relative masses of the molecules written here below the formulae:

$$CH_4 + 2O_2 \xrightarrow{500°C} CO_2 + 2H_2O$$
$$16 + (2 \times 32) \rightarrow 44 + (2 \times 18)$$
$$16 + 64 \rightarrow 44 + 36$$
$$80 \rightarrow 80$$

The units of mass used are, as you will remember, hydrogen atom masses. Hydrogen atom masses are, however, inconveniently small units to measure sensible amounts of materials, and in practice grams or tonnes may be more suitable, depending on the context. Using hydrogen atom units, you can see that the mass of oxygen reacting (64) is related to the mass of methane (16) by a ratio of four to one (64: 16 = 4: 1). In other words, the oxygen mass equals four times the methane mass. Because this is a measurement of proportion it doesn't matter what units are used to measure their relationship. Four times as much oxygen as methane by mass might need to be measured in tonnes if a gas pipe in the street fractured and caught fire. Certainly the quantities involved then would be more than one molecule of methane and two of oxygen. The proportion would remain the same, that is, for every one molecule of methane two of oxygen are used. For every tonne of methane burned, four tonnes of oxygen would be needed.

Because the mass of every methane molecule is the same and remains constant, one gram of methane will always contain the same number of molecules. From the equation we can see that twice that number of molecules of oxygen are used, and they will have four times that mass, 4g. Put another way, there are the same number of molecules in 1g of methane and in 2g of oxygen. This follows from the fact that the molecular mass of oxygen, 32, is twice that of methane, 16. Following this proportional relationship we can see that there will be the same number of molecules in 16g of methane and 32g of oxygen. In this case the number of grams of each substance has been chosen to be the same number as the molecular mass of the compound. The very large number of particles for which this is true is called **Avogadro's number**, after the early chemist Avogadro who first suggested how it could be estimated. The value of Avogadro's number is close to 6×10^{23} (six followed by 23 zeroes).

It follows that 6×10^{23} atoms of hydrogen will have a mass of 1g. As carbon has an atomic mass of 12, so 6×10^{23} atoms of carbon will weigh 12g. That quantity of any compound which is the molecular mass in grams has a special name. It is called a **gram-molecule** or more often just a **mole**.

It is important that you appreciate that a mole of a chemical is a certain number (Avogadro's number) of molecules and hence, because molecules of different compounds have different masses, the moles for different compounds are also different masses. For instance, one mole of $CaCO_3$ has a mass of 100g, and one mole of CaO has a mass of 56g.

To summarise, a balanced equation tells you first of all what are the smallest numbers of molecules of compounds which will react together. This is exactly the same as the number of moles of the compounds which react. Finally, now that you can work out molecular masses (using Table 3.3), you can work out how many grams of substances will react.

Worked example

What weight of quicklime can be slaked with 4.5 litres of water?
The equation for slaking lime (mentioned earlier in section D1) is:

$$CaO + H_2O \rightarrow Ca(OH)_2$$
in molecular masses: $56 + 18 \rightarrow 74$
in moles: 1 mole + 1 mole $\rightarrow$ 1 mole

Thus, if it can be determined how many moles of water are used, it will be known that the same number of moles of quicklime are slaked.

4.5 litres of water = 4.5kg
As 1 mole of water is 18g, 5kg of water is 4500/18 = 250 moles.

From the chemical equation it can be seen that 250 moles of water will slake 250 moles of quicklime. 1 mole of quicklime is 56g, that is, its molecular mass in grams. Therefore, the mass of lime which can be slaked is 250 × 56g = 14,000g or 14kg.

E3   Molar solutions

One circumstance in which you are very likely to meet the idea of the mole is in expressing the strengths of solutions. If one litre of solution contains one mole of compound it is called a **molar solution**. A molar solution at 10% strength consists of 0.1 moles in a litre of solution and would be written 0.1M, or historically as M/10, to describe its strength. The molecular mass of sodium chloride is 22.99 + 35.45 = 58.44, as you can calculate. A solution of sodium chloride at a strength of 58.44g per litre of solution would be a molar solution. Similarly, 5.84g of sodium chloride added to water and made up to one litre would be a 0.1M solution. The strength of a solution expressed in moles is called its **molarity**.

You may well be used to referring to the strengths of solutions (more properly called their concentration) in 'weight per volume' terms, for example, 9.8g/L. This style is often used in publications that discuss recipes for particular conservation treatments, especially for cleaning surfaces. A fuller discussion of this and alternative terminologies is contained in another book in this series, *Cleaning*, but this book demonstrates the advantages of using the **molar concentration** description.

Think about the reaction of sulfuric acid and sodium hydroxide (which is an alkali; note that both solutions are hazardous to handle) to form sodium sulfate and water:

$$H_2SO_4 + 2NaOH \rightarrow Na_2SO_4 + 2\,H_2O.$$

You can see that 1 mole of the acid reacts with 2 moles of sodium hydroxide. When the reaction is complete, and there is no excess acid or alkali, the resulting solution is called neutral (in terms of pH, a term that will be discussed later). Given the molar concentrations of solutions of such compounds you can work out, for example, that a given volume of the acid at a given molar concentration will require exactly twice the volume of sodium hydroxide solution at the same molarity to produce a neutral solution. Or, again, equal volumes of solution will neutralise one another only if the sodium hydroxide is at twice the molarity of the acid. These are examples which further illustrate that a mole is a known number of molecules. Neither example would be obvious in weight per volume terms: that 1 litre of 9.8g/L H_2SO_4 will neutralise 1 litre of 8.0g/l NaOH is not obvious, but that 1 litre of 0.1M sulfuric acid neutralises a litre of 0.2M solution follows from the chemical equation.

4 Atomic structure and chemical bonding

The idea that atoms are able to bond together to form molecules will be quite familiar to you by now. However, so far, nothing has been explained about *why* this should happen. By taking a closer look at the basic structure common to all atoms and then at their electronic properties it will become possible for you to see why atoms of one element behave in a different way from those of any other kind. This chapter explains the several different ways atoms can bond together, helping to reveal how this relates to a variety of physical and chemical properties displayed by the compounds and elements that you work with.

A   The idea of valency

In the last chapter you learnt that the atomic mass of an element is one characteristic feature of its atoms. The structural formulae introduced in Chapter 3 revealed another characteristic; the number of links or bonds which one atom can make with others. The number of bonds that a single atom of an element can make is known as its **valency**. Many elements have single valency, although several metals have two or even more. The equation for the reaction which was used as an example earlier on:

$$CH4 + 2O_2 \xrightarrow{500°C} CO_2 + 2H_2O$$

which describes the burning of methane gas, involves four kinds of molecules with the structures shown in Figure 4.1.

In these structures the number of dashes (which indicate bonds) emerging from each type of atom is seen to be constant. Hydrogen is always joined to another atom by one dash, representing a single bond. Oxygen always makes two bonds to other atoms and carbon always makes four. In order to maintain these characteristic numbers, the structures sometimes show two bonds between one pair of atoms, which is called a double bond.

You must remember, however, that the idea of valency being constant for every atom is only a *concept* and it is not totally consistent (carbon monoxide is one example when it does not appear to work). However, for a wide range of compounds, especially organic ones, it is extremely useful and it gives insight into chemical reactions. Table 4.1 provides the valency of several elements commonly found in organic molecules.

DOI: 10.4324/9781003261865-4

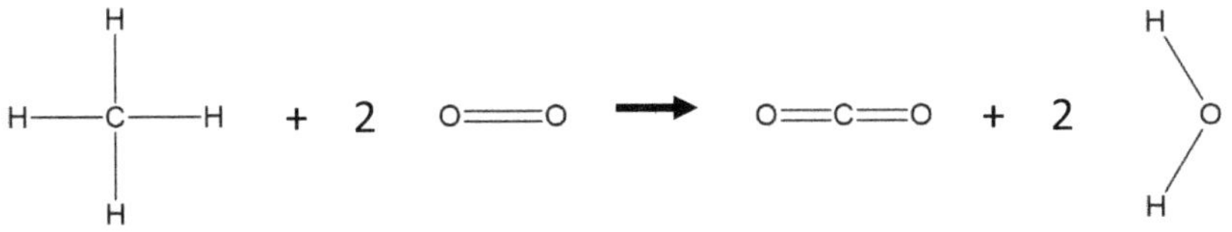

Figure 4.1 Structural formulae for molecules involved in the burning of methane.

Table 4.1 Valencies of some common elements.

Valency	Element			
1	H	F	Cl	Br
	hydrogen	fluorine	chlorine	bromine
2	O	S		
	oxygen	sulfur		
3	N	P		
	nitrogen	phosphorus		
4	C			
	carbon			

Exercises

1   Check the structural formulae of ethyl alcohol (see Figure 3.6(a), dimethyl ether (see Figure 3.6(b)), ethylene oxide (see Figure 3.7(a)) and acetaldehyde (see Figure 3.7(b)) to confirm that the correct valencies have been used.

2   Draw structural formulae for the following molecules:
   (a) H_2
   (b) N_2
   (c) NH_3 (ammonia)
   (d) CH_5N (methylamine)
   (e) C_2H_3Cl (vinyl chloride)
   (f) the third isomer of C_2H_4O.

Clearly the restrictions on how many bonds each kind of atom will make are the major influence in determining what proportions of different elements appear in any given compound. To understand the origin of the bonds and see why particular valencies occur for a given element, and to discover the limitations of this simple pattern-making exercise, more needs to be understood about what atoms themselves are made of.

B   The structure of atoms

B1   *The electrical connection*

You will, no doubt, be familiar with the effects of static electricity when polishing a glass case or brushing one's hair or even stroking a cat in dry, frosty weather. As you polish, forces of attraction are built up between the case and the cloth, so that dust is often attracted to the surface you are trying to clean. These forces of attraction and repulsion are fundamental to the nature of electricity. The **electrostatic forces** produced by friction have been studied in great detail and it has been established that, as with magnetic forces, some attract and others repel. Objects are able to build up different degrees of static electricity; this is described as **electric charge**. There are two kinds of electric charge. Objects with different (opposite) charges tend to attract one another, while those with the same charge repel one another. It has also been observed that these forces of attraction and repulsion between charged materials rapidly decrease as the objects are moved further apart.

   Further research has established that atoms themselves are made up of charged particles, some carrying charge of one kind and some of the other. These two types are denoted by the terms **positive (+ or +ve)** and **negative (− or −ve)**. When charging through friction takes place, a few of these minute particles become transferred from one atom to another, so that the normal, even mixture of positive and negative charges in a single atom is distorted. A dominance of one or other type is created, causing an overall positive or negative charge to build up on the rubbed surface.

B2   *The composition of atoms*

There are two types of **charged sub-atomic particles** in atoms:

Proton	which carry a positive electric charge and have almost the same mass as a hydrogen atom.
Electrons	which carry an electric charge of exactly the same 'strength' as that on the protons, only negative. They are very much lighter, only 1/2000th the mass of a hydrogen atom.

   Atoms also contain a third type of sub-atomic particle:

Neutrons	so called because they are electrically neutral (meaning, they carry no charge). They have almost the same mass as protons (in fact they are very slightly heavier).

In the past the structure of atoms has been compared to our solar system, which includes the sun and planets. At the centre of an atom is a **nucleus** (the plural is '**nuclei**') which contains only the heavy sub-atomic particles (the protons and neutrons). Just as the sun contains almost all the mass in the solar system, so the nucleus of an atom contains almost all the mass of the atom.

Individual atoms (helium, oxygen, neon, etc.) are electrically neutral and so for every proton in the nucleus there must be a balancing negative electric charge. This is provided by the electrons which surround the nucleus. There is the same number of electrons as there are protons. The electrons are in constant motion, effectively in orbit around the nucleus at the centre, which can be paralleled by the motion of planets around the sun. The sun attracts the planets in orbit round it by gravitational force and, in a comparable way, the nucleus holds on to its electrons by an *electrostatic* force. This force is attractive because electrons and protons have opposite charges.

The analogy with the solar system cannot be taken very far. (For example, there is no equivalent to the Earth's moon in atoms.) In an atom, the electrons are moving very fast and the whole system is extremely small, even allowing for the very wide empty space between the nucleus and the orbiting electrons. It is not possible to say precisely where each electron is at any one time, and not just because they are too small to 'image' with almost all instruments or microscopes. It is much more realistic, therefore, to think of a fuzzy, roughly spherical volume of space surrounding the nucleus in which all the electrons must be found though they are not precisely located as dots on its surface. This is often described as the **electron cloud**.

This model of an atom can help you to understand much more about the behaviour of atoms. It suggests, for example, that the chemistry of an atom is connected more with the outer electron cloud than the nucleus. It is the electron clouds which form the outer parts of atoms and, consequently, come into contact with one another when atoms collide. It is also in accord with the visual model of a molecule given at the beginning of Chapter 3 where it was suggested that atoms, joined together to form a molecule, merge into each other. This is, of course, easy to imagine if the outer parts of the atom are cloud-like rather than rigid. However, this visualisation of the atom will have to be developed further still to explain why colliding atoms sometimes bond together to make molecules, but on other occasions bounce apart.

Although there are well over a hundred distinct types of atoms (the elements), they are all composed of these three sub-atomic particles. The difference between atoms is in the number of protons present in their nuclei.

B3   *The atomic nucleus, mass numbers and isotopes*

It has been emphasised just how invisibly small whole atoms are, but the nucleus of an atom is small even in comparison with that atom – less than one thousandth of the atomic diameter. Into this tiny volume are packed all the protons (which are the particles with positive electric charges). The forces of repulsion between them are very great and so there has to be some other force holding them together. The neutrons are involved here, and (except for hydrogen which has a nucleus with only one proton) all atomic nuclei contain neutrons, at least as many as there are protons, in every case.

All elements are chemically different from the other elements and this is because the chemical behaviour of an element is governed by the number of electrons that surround its nucleus. For the electric charges within an atom to balance there must be the same number of electrons in orbit outside as there are protons inside the nucleus. Consequently, the feature which distinguishes atoms of one element from another is the *number of protons* in the nucleus. Each distinct element has a specific number of protons. Hydrogen, the lightest atom has only one; helium, the next in mass, has two; lithium has three and so on, up to the heaviest naturally occurring element, uranium, which has 92, and then beyond. The number of protons in an atom's nucleus is called its **atomic number** (whose symbol is **Z**) and this provides a specific and systematic way of identifying an element.

The number of neutrons in the nucleus is about the same as the number of protons, although the proportion of neutrons increases in the heavier elements. Hydrogen does not have any neutrons; in helium there are two protons and two neutrons; in aluminium there are 13 protons and 14 neutrons; in the heavy element, gold, there are 79 protons and 118 neutrons. (This is all referring to the lightest isotope, which is in many cases the most abundant stable isotope.)

The number of protons is constant and specific for any given element. However, for many elements, the nuclei of atoms of that particular element do not always have the same number of neutrons. For example, the nuclei of atoms of copper which all contain 29 protons, can have either 34 or 36 neutrons in them, which means there are two isotopes for copper. Since the electrons in an atom are so very small and light compared with the nucleus, and since protons and neutrons have masses very close to one 'hydrogen atom unit', (Chapter 3, section E) the total mass of the atom can be fairly accurately described by adding the number of protons to the number of neutrons. This gives a whole number called the **mass number**. The two isotopes of copper thus have mass numbers of 63 and 65 ($29 + 34 = 63$ and $29 + 36 = 65$). The copper in copper ore occurs as a mixture of its two isotopes; the lighter isotope is more than twice as common as the heavier. In a sample of copper metal or in a copper corrosion product, the two kinds of atom will be completely mixed and there is no way of separating them chemically. If calculations related to chemical reactions are used to determine the atomic mass of copper, the answer will come out as an average between the two. In tables of accurate atomic mass that of copper is listed as 63.54 whereas any individual copper atom will have an atomic mass which is a whole number, either 63 or 65.

When this is written down, isotopes are distinguished by first adding an isotope number in superscript to the symbol for that element. The isotope of carbon with six protons and six neutrons is thus written as ^{12}C, but referred to in speech as 'carbon-12'. The isotope of carbon with six protons and eight neutrons is written as ^{14}C. This heavier isotope is formed in the atmosphere from nitrogen (which is an abundant gas) interacting with a cosmic ray. The ^{14}C soon reacts with oxygen to form $^{14}CO_2$. ^{14}C is an unstable isotope and over long periods of time it emits radiation and decays back to nitrogen whatever compound it is in: this isotope is known as carbon-14 or radiocarbon, meaning 'radioactive carbon'. Plants absorb CO_2 as they grow, and thus incorporate some ^{14}C into their structure, the proportion to ^{12}C matching what was in the atmosphere that year. Thus annual crops that are harvested in the same year of planting, like grain and the flax or cotton used to make linen or cotton canvas respectively, have a well-defined year of origin, which has some potential in the period from the 1950s to about the end of the 1990s for measuring the proportion of ^{14}C in canvas made in these decades, when nuclear weapons testing increased the concentration of atmospheric ^{14}C over the typical level for that century. It is possible, using an independent calibration curve, to infer whether a canvas was made in one of these decades rather than before or after that period, sometimes to within a period of a few years. The time for half of the ^{14}C atoms to decay is very long; 5730 years, which is called the half-life for this isotope. Most trees live for a tiny fraction of this time, and measurement of the declining concentration of ^{14}C in wooden posts excavated from an archaeological site or timbers used in an old building can be compared to another independent calibration curve to give a date corresponding to the period when the tree was alive and growing. If during conservation treatment a carbon-containing material is introduced into an object that is made of an organic material, for example a polymer is used as a consolidant for fragile archaeological wood, the ratio of the ^{14}C and ^{12}C isotopes which is characteristic of the date when the original plant material was alive will be distorted, because the new material will have a different radiocarbon age (if the new polymer is derived from petrochemicals that are themselves derived from plants, it will be a much older radiocarbon age) and thus a different **isotope ratio** from the archaeological wood. As a conservator, you should be aware that **archaeometric evidence,** or occasionally evidence useful for suggesting whether a canvas is consistent with the lifetime of a particular artist, would thus be destroyed by a consolidation treatment or application of an irreversible coating or any organic material, while application of any organic adhesive would affect a more localised area. In many circumstances, this is not sufficiently concerning to prevent treatment that would ensure the survival of an object.

B4   *The electronic structure of atoms*

Early attempts to describe what goes on within the electron cloud visualised electrons as circling the nucleus, in orbits that are predictable over very long periods of time, just as planets progress around their sun. It is now known that it is impossible to measure accurately what the position, speed and direction of travel of an electron are, all at the same time. All one can hope to do is to specify a volume of space around the nucleus and say that there is a high chance of finding the electron in that space, if it were possible to go 'looking' for it. These volumes of space are called **orbitals**. There are orbitals around the nucleus of every atom. Each orbital can contain two electrons.

Orbitals have many different shapes; those of the heavier elements are quite complicated because they are so 'busy' with electrons each seeking to maximise its distance from the others. These complex orbital shapes penetrate into one another. Despite this complexity, the averaged effect is that the greatest density of electrons is found in concentric spherical *shells* surrounding the nucleus, and in diagrams they are very often simplified to the shape of a sphere. This **shell structure** is shown in Figure 4.2 in two different forms.

It is possible to define these shells in more detail, because each comprises orbitals, each of which can accommodate two electrons. These orbitals are of four different types, termed 's', 'p', 'd' and 'f', which have different shapes. The first shell with only two electrons has only a single s orbital, and has the designation '1s'. The second shell contains s and p orbitals termed 2s and 2p respectively. There are three 2p orbitals, each with the same shape, but oriented differently, so there are eight electrons in this second shell. The innermost shells will nearly always be completely filled by electrons, because having them filled is a stable configuration. The outermost shell may be full, nearly full, half empty or almost empty. It is these four rough categories of 'filled-ness' which determine the chemical character of an element.

Atoms in which the outermost shell of orbitals is completely full are very stable. Of particular significance in determining chemical behaviour is a **filled outer shell** containing eight electrons – a condition which many atoms can achieve through adding or losing an electron or a few electrons in the course of a chemical reaction. There are five elements whose atoms do not enter into any chemical reactions. These atoms have totally filled outermost shells and can collide with other atoms and molecules without reacting with them, so they do not form compounds with other elements. Nor do they form bonds with other atoms of the same element as do, for example, oxygen and nitrogen. These elements are neon, argon, krypton, xenon and radon, which are gases found in the atmosphere in small amounts, known as **inert gases**, and formerly as the noble gases.

The principle of **bonding** is that atoms can share or swap electrons between outer shells in order to achieve the stable condition of eight electrons in their outermost shells. Atoms involved in these exchanges then stay close together, forming molecules. (Hydrogen molecules H_2 and helium atoms, however, do not fit this pattern of eight, as the first orbital shell – the shell which is closest to the nucleus – is filled by the only two electrons that they have. It is the 'filled-ness' of that lone shell that is the important point here.)

(a)   (b)

Figure 4.2 (a) Representation of the electron cloud in two dimensions, showing three 'shells' with the greatest electron density; (b) Three-dimensional representation showing the shells nested inside each other.

To understand chemistry you have to look further at the way in which atoms acquire stability. They do this through 'giving', 'taking' or 'sharing' the electrons occupying their outermost shells with other atoms, in order to achieve a full complement of eight electrons in their outer shells. Which of these methods is used depends on how many electrons the atom has in its outermost shell to begin with. Atoms which have a nearly empty shell of electrons will readily give electrons away to expose the underlying filled shell. Atoms with a nearly full shell will readily accept electrons to make up the full number of eight in the outside shell. In between, the atoms with half-filled outer shells will tend to share electrons. These properties can be illustrated most clearly by considering the 18 lightest elements, shown in Table 4.2.

If these elements are grouped according to how many electrons they have in their outermost shells (and hydrogen is omitted because of its exceptional structure), one starts to create a **periodic table**, as shown in one form in Table 4.2 for the first 18 elements, and at the end of this book in its standardised form for easy reference. The elements in each column of the standardised table not only have the same number of electrons in the outer shell but also share somewhat similar chemical properties. This pattern was recognised in the nineteenth century, long before anyone could explain it in terms of the electronic structure of atoms. You will become aware of some of the similarities amongst elements and their compounds which Table 4.2 shows.

For elements heavier than calcium the regularity of this periodic pattern is distorted. Look at the periodic table at the back of this book. Although it is not regular,

Table 4.2. The first 18 elements, demonstrating the degree to which the electron shells are filled.

Number of electrons in atom	Name of element	Symbol for the element	Notes
1	hydrogen	H	1st shell not filled
2	helium	He	1st shell full
2	lithium	Li	2nd shell has only 1 electron
4	beryllium	Be	2nd shell has 2 electrons
5	boron	B	2nd shell has 3 electrons
6	carbon	C	2nd shell has 4 electrons
7	nitrogen	N	2nd shell is short of 3 electrons
8	oxygen	O	2nd shell is short of 2 electrons
9	fluorine	F	2nd shell is short of only 1 electron
10	neon	Ne	1st and 2nd shells full
11	sodium	Na	3rd shell has only 1 electron
12	magnesium	Mg	3rd has 2 electrons
13	aluminium	Al	3rd shell has 3 electrons
14	silicon	Si	3rd shell has 4 electrons
15	phosphorus	P	3rd shell is short of 3 electrons
16	sulfur	Si	3rd shell is short of two electrons
17	chlorine	Cl	3rd shell is short of only 1 electron
18	argon	Ar	1st, 2nd and 3rd shells all full

the full periodic table can still be explained systematically in terms of the numbers of electrons in different types of orbital. To explain the mechanisms of valency, without too much complication, only the first two rows of the table will be used here. The ways in which bonds are formed vary according to which of these classes of atoms are brought together. Several types of combination will now be looked at in turn.

C Bonding mechanisms

C1 Covalency: bonding by sharing

Methane is a molecule containing two elements that both have atoms possessing half-full outermost electron shells. It is an example of a molecule made by atoms sharing electrons. Bonds formed by this mechanism are called **covalent bonds**. Figure 4.3 represents the atoms of hydrogen and carbon diagrammatically, showing the electron shells and their electrons.

Figure 4.4 shows a model of the methane molecule, demonstrating the origin of the electrons forming the bonds between the atoms by using the same symbols. The first shell of each hydrogen atom is made complete by sharing one electron from the carbon. The first shell of the carbon atom is already full. The second shell of carbon is made up to its full complement of eight electrons by sharing one electron from each hydrogen atom. You will notice that the electrons in the two atoms have been represented by two differently coloured symbols. This has been done purely to illustrate the origin of the electrons when full shells have been formed by sharing. It is important to remember, however, that all electrons are in reality identical.

These models are not drawn in exactly the same way in every book you read but you will always be able to see shared electrons. A common form for diagrams such as these suppresses the superfluous information about the inner full shells and presents methane as shown in Figure 4.5(a), as a simplified two-dimensional molecule.

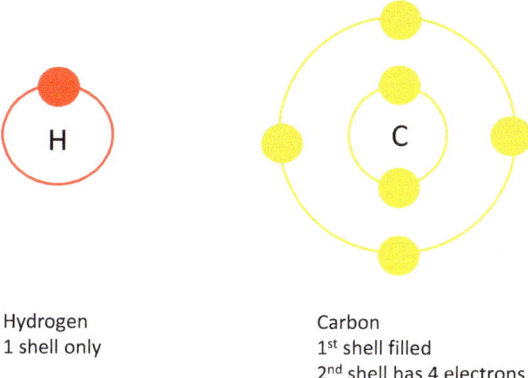

Hydrogen
1 shell only

Carbon
1st shell filled
2nd shell has 4 electrons

Figure 4.3 Electronic structure of hydrogen and carbon, shown in two dimensions.

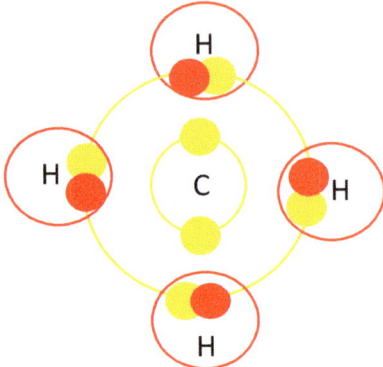

Figure 4.4 Full electronic structure of methane, shown in two dimensions with contributory electrons.

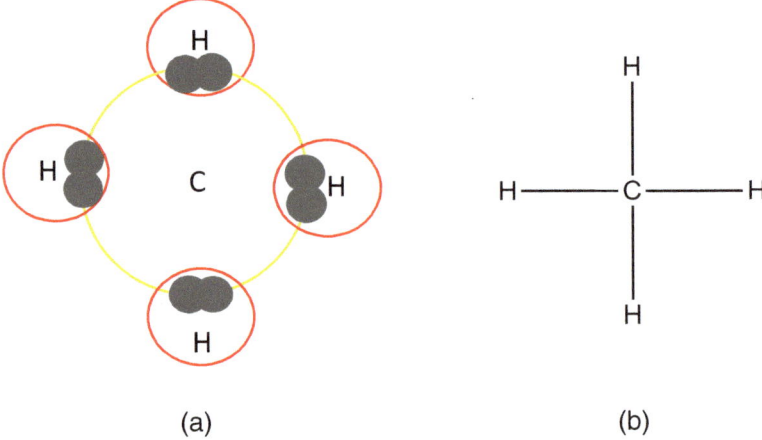

(a)   (b)

Figure 4.5 (a) Simplified two-dimensional structure for methane; (b) Simplified structural formula for methane.

Here the rings are used merely to enclose electrons shared by each atom. Once the molecule is formed, the origin of any individual electron is unimportant and cannot be determined by measurement, so distinct symbols are quite meaningless. The shared electrons are therefore all assumed in the simpler diagram in Figure 4.5(b). It should by now be obvious why the valency of carbon is always four and the valency of hydrogen is always one.

You can see that all five atoms in Figure 4.5(a) have achieved full outer shells, and consequently there is no tendency for any more electron swapping or sharing to occur. The whole group 'looks' rather like the inert gas neon with the full complement of eight electrons in the outer shell. As a result, these molecules can endure rapid movement and collisions without undergoing chemical change. Only a collision which is violent enough to break the shared bonds will provoke a chemical reaction. If you compare the new picture of methane, Figure 4.5(a), with its structural formula in Figure 4.5(b), you can see that the bond shown by one dash means the sharing of a pair of electrons between the atoms. The covalent bond is called an **electron-pair bond**.

Electron models corresponding to other structural formulae can be drawn. Two are shown in Figure 4.6. In Figure 4.6(b) you should pay attention to the 'double' bond between the carbon atoms which represents *two* pairs of electrons being shared by two atoms.

By now, the reader already studying the conservation of organic materials generally must have noticed that a great number of solvents and organic reagents used to remove soiling from surfaces sound as if they will fit into this category of materials that exhibit covalent bonding. In fact, a great number of organic compounds fit too. Covalent bonds are relatively weak, and this is why such materials are often liquids at room temperature, and why the ones that are liquids are often rather volatile.

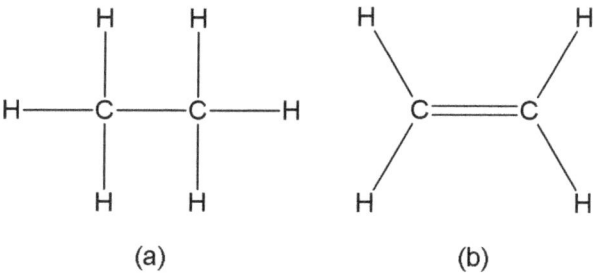

(a)   (b)

Figure 4.6 Structural formulae and electron-sharing diagrams for two more compounds of carbon and hydrogen: (a) the gas ethane and (b) ethylene, more correctly called ethene, which is the building block for polythene.

Exercises

3 Construct electron-sharing pictures for

   (a) ammonia, NH_3
   (b) nitrogen, N_2 and
   (c) methylamine, CH_3NH_2.

To do this you will first have to construct an electron model for a nitrogen atom. Consult the partial periodic table above (Table 4.2) to remind yourself how many electrons a nitrogen atom has. Check that in your complete molecules each H atom has a share of two electrons and other types of atoms have shares of eight electrons.

Another value of these models is that they show that each atom in a molecule is 'satisfied' with its complement of electrons without any atom receiving or losing any electrical charge. This fact means that the molecules have little tendency to stick together, having become almost like the atoms of the inert gases. The consequence is that covalently bonded molecular compounds, unless of high molecular weight, *tend* to have low melting points, and to form mobile liquids or weak, easily distorted solids (plastics for example). You must note, however, that the word 'tend' denotes a range of behaviour that is very, very wide.

Electron takers in covalent bonds

When you studied the reaction of methane with oxygen you noted that oxygen always occurs in 'oxygen molecules', and not as single oxygen atoms. However, oxygen would rather take electrons than share them. If there are no other atoms nearby which *give* electrons, one oxygen atom will share electrons with another, forming an oxygen molecule. Covalent sharing is the mutually satisfying answer and oxygen molecules form as in Figure 4.7 where the electrons of the two atoms are distinguished to show what is going on. You can see that according to the pattern of bonding so far developed, this must be a double bond as in Figure 4.1.

Figure 4.8 is a representation of the arrangement of electrons in a water molecule. This appears as straightforward covalency, but the oxygen atom is so demanding of electrons that the molecule is most stable when the oxygen atom attracts the hydrogen atoms' electrons towards itself. The positive hydrogen nuclei (protons) are therefore exposed and the oxygen atom has a slight negative charge because of the

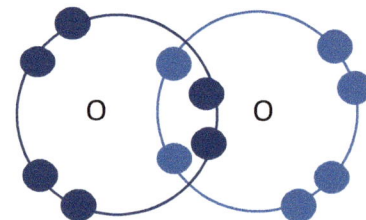

Figure 4.7 Structural representation of an oxygen molecule O_2.

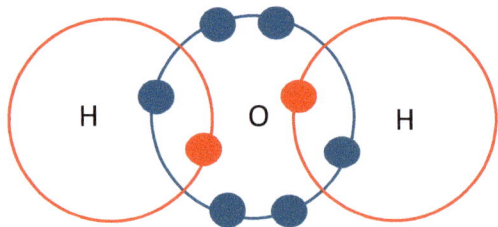

Figure 4.8 Structural representation of a water molecule H_2O.

extra share of negative electrons. (Remember that electrons are not point charges in their orbitals but rather they are clouds of negative charge.) This slight charge imbalance is the key factor in the useful solution-making properties that water has, and is critical for its importance in cleaning, especially when it is combined with other materials. An `added important factor is that the directions of the bonds the oxygen atom will make are not diametrically opposite but are at slightly more than a right angle (108° from both measurement and calculation) to each other. This is because, as with the methane molecule described earlier, the negatively charged bonding and non-bonding electron pairs arrange themselves to be as far apart as possible – at the corners of a tetrahedron (which is a four-sided, three-dimensional shape), but the nucleus of the oxygen molecule, with its 16 protons, is attracting them strongly compared to the pull of the lighter hydrogen nuclei. A somewhat more realistic representation of the water molecule is therefore like that shown in Figure 4.9.

The similarity to the three-dimensional structure of methane (Figure 4.10) should be obvious.

A molecule which has two separate but opposite charged regions is called **polar**. The **polarity** of water, which is due to the non-symmetrical arrangement of the bonds between its atoms, has a profound effect upon its properties. Indeed, water has some very peculiar properties, physical properties as well as chemical ones, that are not generally seen in other molecules with such a simple structural formula. For example, water has a very high freezing point at 0°C and a high boiling point at 100°C at normal atmospheric pressure, compared to the boiling point of methane at -161.6°C, which ensures it is always encountered as a gas. Water's

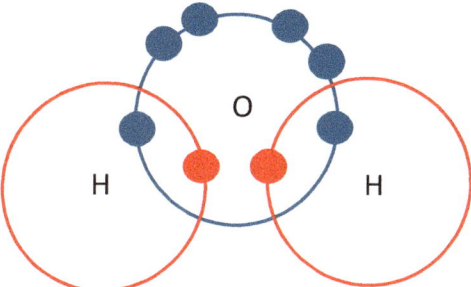

Figure 4.9 More realistic structural representation of a water molecule H_2O.

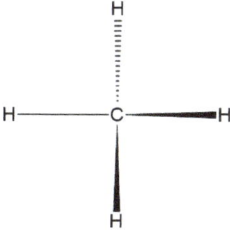

Figure 4.10 Structural representation of a methane molecule.

polarity is ultimately the reason why we inhabit such a watery planet and why we ourselves include a lot of water, well over half our body weight. The peculiar properties of water are in fact responsible for the development of life on this planet: remarkably few other compounds could be substituted for it. Water and its properties make it a unique reagent to be used by conservators too.

Covalent bonds which are 'dative'

It is also possible for atoms to make sharing bonds in which both the electrons involved in a bond originate from only one of the atoms. Such a bond used to be known as a **dative bond**, meaning a 'giving' bond, though today it is usually described as a **covalent bond**, and even more correctly as a **coordinate covalent bond**.

Sulfur dioxide is a pollutant generated by the burning of fossil fuels, which causes acid attack on masonry and on leather book binding, and which can react with true fresco paintings and cause deterioration. Its molecular formula is SO_2. Both kinds of atoms in the molecule have outer shells containing six electrons. The sulfur atom is known to lie between the oxygen atoms. Figure 4.11 shows a possible electronic structure which obeys the 'rule of eight' (having an outer 'complete' shell of eight electrons), where the bond labelled 1 is an ordinary double covalent bond exactly mirroring the central bond in ethylene (see Figure 4.6(b)), but the bond marked 2 is different. To make the model of the rule of eight work, both the electrons in this bond have to be described as coming from the sulfur atom, which is why it is called a *coordinate* covalent bond.

Molecular orbitals

However, experiment has shown that the bonds between the oxygen and sulfur in sulfur dioxide are identical, just as one might guess they would be for the sake of symmetry, and not unequal ones made from one single and one double bond as the model would suggest. Figure 4.12 is another attempt to convey the arrangement:

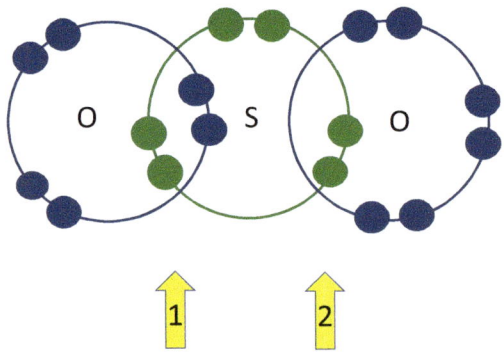

Figure 4.11 An attempt to make the 'rule of eight electrons' work for SO_2.

the electrons shown as open circles could be visualised as moving in their orbits to create a more balanced arrangement. Thus, in more nuanced theories, the bonding of molecules such as SO_2 is explained more accurately by considering the electron cloud in terms of the whole molecule: such a concept leads to the term 'resonance structure' for such a molecule. Just as there are *atomic orbitals* which are stable zones for electrons surrounding the one nucleus of a single atom, so there are **molecular orbitals** which are stable electron arrangements surrounding the several nuclei of a molecule.

Similar molecular orbital diagrams could be drawn for all the small molecules that we have considered, such as oxygen and methane. Molecular orbitals can be thought of as being formed by the overlapping of the atomic orbitals. The eight-electron rule does not accurately predict all the properties of simple molecules. However, the mathematics involved in the more complete molecular orbital theory is very complicated and it cannot provide so simple and useful a guide to molecular structure as the rule of eight.

C2 *Ions: bonding by electron transfer*

The periodic table shows that the atoms of the lighter metals (Li lithium, Na sodium, and K potassium are in this category) have only a few electrons in excess of a stable inert gas structure – just one or two (in these cases, one each for Li, Na and K). The basis of the **ionic bonding** system is that metal atoms can acquire the stable outer shell of eight electrons by in effect completely shedding those few extra electrons. Of course, there must be somewhere for the electrons to go. You may well have deduced that the lost electrons go into the outer orbitals of atoms which need to *accept* a few electrons in order to achieve the stable eight-electron configuration – which is in a sense extreme electron sharing. Oxygen and chlorine are two common elements in this situation, and the compounds which are formed when such an exchange of electrons has occurred are known respectively as **oxides** and **chlorides**.

The atoms involved in these exchanges become electrically charged because an imbalance in the numbers of electrons and protons in the original atom is created through loss or gain of electrons. The charged atoms are known as **ions**. The elements which readily lose electrons (metals) are called **electropositive** elements

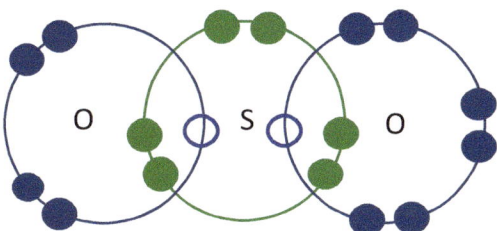

Figure 4.12 Another attempt to visualise the structure of SO_2 in two dimensions.

because the ions they form have a positive charge – the result of shedding negative electrons. This ionic bonding is best described by the electron pair in a covalent bond being displaced so far towards the more electronegative atom that the electron pair sits inside the outer shell of that atom. If put in water the polar nature of the water molecule stretches the 'ionic' bond until water molecules can fully surround (solvate, that is dissolve) it.

The loss of electrons can be written like a chemical equation in the following way:

$$Ca - 2e^- \rightarrow Ca^{2+}$$

Calcium atom minus two electrons becomes a calcium ion with two units of charge, usually written as '2+' but in the past also as '++'.

Positive ions are also known as **cations** (pronounced as *cat-ions*) because during electrolysis they migrate to the negative electrode known as the **cathode** (pronounced *cath-ode*, the end connected to the negative pole of the battery which, of course, attracts positive particles).

Negative ions, known as **anions** (pronounced as *an-ions*) (which are attracted to the positive electrode known as the **anode** during electrolysis) are formed from **electronegative atoms** which collect the electrons lost by metal atoms. In an equation the change can be written as a reaction between an atom and one or more electrons:

$Cl + e^- \rightarrow Cl^-$
Chlorine atom plus an electron $\rightarrow$ –vely charged chloride ion
$O + 2e^- \rightarrow O^{2-}$
Oxygen atom plus two electrons $\rightarrow$ oxide ion carrying 2 units of charge

In these equations you will have noticed ions written as the atomic symbols carrying + and – signs. The convention is that the number of electrons lost or gained by an atom in becoming an ion is shown as a **superscript** to the right of the element symbol with a plus or minus next to it to show the residual charge. Thus a chloride ion, Cl^-, is a chlorine atom which has accepted one electron, while the oxide ion O^{2-} carries two extra electrons. Similarly, the copper ion Cu^+ is a copper atom which has lost one electron (once commonly called a cuprous ion but now often written as a $Cu(I)$ ion because there is one electron involved), while another condition of ionised copper, Cu^{2+}, is *two* electrons short of being a complete atom (once commonly called a cupric ion but now written as a $Cu(II)$ ion, where the 'II' is a Roman numeral). By analogy with the number of bonds an atom forms in a covalent compound, the number of charges on an atom is called its **ionic valency**. So Cu^{2+} is called two-valent or more usually **di-valent**. These names are further explained in the next chapter.

Ions need not only be single atoms with an excess or deficit of charge. Anions in particular occur as clusters of atoms which are as stable as molecules, but are charged. You will meet many examples in the following chapter, but among compounds already mentioned lead white pigment contains OH^- ions (the hydroxide portion) and CO_3^{2-} ions (the carbonate portion). Gesso, calcium sulfate, contains SO_4^{2-} ions.

Once ions of opposite charge are formed by the transfer of electrons from one atom to another there will be an electrostatic force of attraction between them. It is this force which constitutes the **ionic bond**. Moreover, because of the non-directional nature of electrostatic forces it becomes meaningless to talk of a 'mol-ecule' of an ionic compound. In the first section of this chapter the number of bonds one atom could make to others (valency) was mentioned and it was seen in discuss-ing the mechanism of covalency how that number was limited. Charged ions of one sign, in contrast, will attract *any* ions of opposite sign, and from any direction, causing every ion to be surrounded by oppositely charged ions. Equally, ions of the same sign repel one another, and thus structures develop where these attractive and repulsive forces are so organised that there is a net attractive force. As a result, ionic solids form **crystals** whose regular shapes reflect an orderly arrangement of positive and negative ions, each surrounded by one of the opposite charge. More will be said about crystals in section D of this chapter, but meanwhile let us look at what has happened to the idea of molecules in such a compound.

If quicklime is made (CaO, which is very reactive and must be handled with care, using eye protection and alkali-resistant gloves) by roasting limestone, the result is a white powder. Each particle of the powder must contain billions of ions of two kinds, Ca^{2+} and O^{2-}, in equal numbers. Each calcium atom has two electrons to give and each oxygen atom will accept two. In three dimensions the assembly is a crystal; the diagram (Figure 4.13) shows a representation of this.

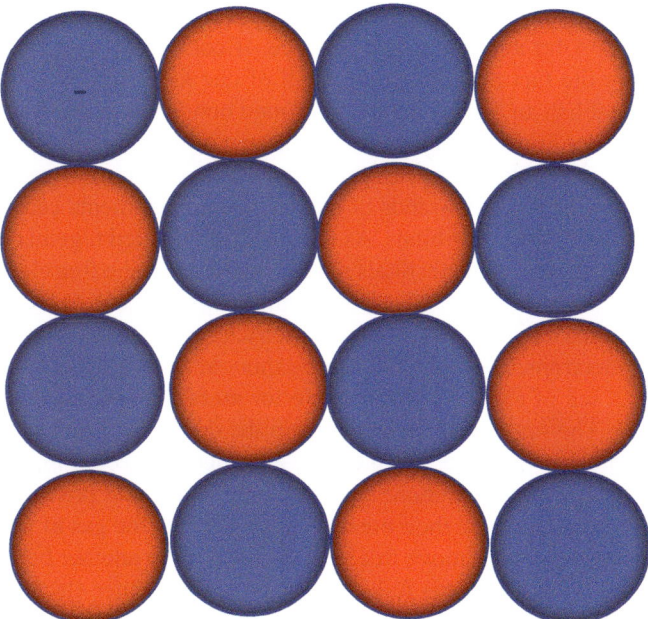

Figure 4.13 A two-dimensional representation of the alternation of positive and negative ions.

Each ion is surrounded by a regular and fixed number of oppositely charged ions but is not associated with any one ion in particular. Each little crystal has grown from an embryo of just a few ions but it is most unlikely that every crystal will be the same size. Out at the 'ragged' edges of each crystal there is less stability. The formula of an ionic compound such as CaO means something different from the formula for a covalent one. With a covalent compound the formula shows the particular grouping of atoms which stick together as a molecule but for an ionic compound it merely indicates the relative proportions of elements which exist in the substance.

Ionic bonding is common in inorganic materials. And because ionic bonds are usually much stronger than covalent bonds, such materials are almost all solids at room temperature, and almost all of them have very high melting points too. Artefacts that can readily survive for thousands of years rather than tens or hundreds of years tend to have bonds of this type. Ceramics, stone and many of the traditional pigments used since antiquity fall into this group, but since these classes have different physical properties, clearly they have different types of bonding too. Coloured materials like pigments are so complicated and interesting that they deserve a different book to do justice to the role that valency can play in creating colour, so they won't be discussed any more here.

C3   Bonding in metals

Most metals, as you know, are strong and solid materials, and often lustrous (shiny) in appearance at ordinary temperatures. (Exceptions such as mercury that is a liquid at room temperature are rarely encountered, largely for health and safety reasons, and it is possible that the reader thought at this point that all metals could be described as solid and lustrous when free of corrosion.) We have to find an explanation for the fact that these atoms, all of one kind, which all want to lose electrons to establish an inert structure, are found bonded together. If electrons can be distributed over more than two atoms to form a bond then it is not too difficult to suppose that 'free' electrons, also known as valence electrons, can collectively be shared equally among thousands of atoms. Metallic elements carry only a few electrons in their outermost orbitals. As the atoms are packed together into a solid, these sparsely populated outer orbitals readily overlap and the free electrons which they contain cease to 'belong' to any particular atom and are described as being de-localised. Since a pure metal conducts electricity by the coordinated movement of these free electrons, the electrons can be thought of collectively as something like a gas, or a sea of electrons, responsible for **metallic bonding**, as represented in Figure 4.14. The same concept can explain the ductility of metals, their ability to deform before they fracture, and it can be extended to account for the formation of metal alloys, where two or more types of metallic element provide the positive ions, and all types contribute their outermost electrons to the gas of free ones.

These electrons can thus move freely from one atom to another because none of the overlapping outer orbitals is full. In other words, the outer electrons belong to all the atoms. The mobile electrons act as a cohesive force preventing the positive metal ions from pushing each other apart. Just as with the ionic bond there is no distinct group of atoms which can be identified as a molecule. This type of structure

Figure 4.14 Electron sharing among several metal atoms to create a metallic bond.

can extend indefinitely in any direction. The electrons as a group can 'step' from one metal nucleus to another without changing the structure, which is what creates electrical conductivity. Often the most stable structure (and hence the most likely one) will be that in which there is the greatest overlap of the orbitals of one atom with those of its neighbours. This is achieved in many metals (such as copper, silver and gold) by a regular pattern called **close packing**. The regular array of repeating units in three dimensions suggests that metals are **crystalline**. It is unusual to see individual metal crystals in isolation, but solid metal objects are made up of large numbers of small but interlocking crystals. The boundaries between these crystals that contain multiple thousands of atoms can be seen when a suitably polished and prepared cross-section of the metal is viewed under an optical microscope.[1]

Three types of bond?

Covalent, ionic and metallic bonds have been described as if they formed three distinct categories. We have seen that really these three are just definable points in a wide range of bonding behaviour. These three types of bond are a good starting point when degradation processes and their chemistry need to be investigated. The extreme form of this is the completely de-localised electron structure of metals.

In a covalent bond the electrons may not be shared equally between the atoms if one is more electronegative than another (for example, oxygen in H_2O). The most extreme form of unequal sharing is when there is complete electron transfer from one atom to another, as is found in the ionic bond.

D   Physical properties related to bonding

The materials of which objects are made have different physical characteristics which distinguish them from each other. Materials are often chosen for a particular application because of these distinguishing features. Copper, which conducts electricity extremely well, is used to form connections in computer chips, while

---

1 Examples can be seen in Struers Ltd. 2022. *Application Notes: Metallographic Preparation of Cast Iron*, as Figure 4, and the figure labelled 'ferritic malleable iron' that has no number, downloaded from https://www.struers.com/en/Material.

plastics such as polyvinyl chloride, which do not conduct electricity, are used as insulation for copper wire used for domestic electric cables, and as the substate for computer chips. Solvents with low boiling points, covalent bonds, and chlorine substituted for some of the hydrogen atoms found in more usual conservation solvents are used for commercial dry-cleaning, because they will evaporate fairly rapidly from the textile, once the cleaning has been finished.

The range of possible properties is very great. Materials may be heavy, light, opaque, transparent, volatile, involatile, rigid, fluid, electrically conductive, or non-conductive. Many of these properties can be directly related to the kind of bonding within the material. If you ask four questions about the physical properties of a particular material, you will find that it can be placed in one of four classes, which, broadly speaking, indicate four different types of bonding and structure.

The questions are:

Are the temperatures at which the material melts and boils low or high? (A convenient dividing line is 200°C for the boiling point. Many substances will decompose above this temperature rather than melt or boil, but even so they still belong in the high temperature class since they do not change state at a temperature below 200°C.)
Can the material be made to conduct electricity freely?
Does the material conduct electricity in both the solid and the liquid states?
Does it only conduct electricity in the liquid state, that is when molten, or when dissolved in a liquid?

The first two groups labelled type 1 and type 2 in Figure 4.15 are covalently bonded materials while type 3 and type 4 have metallic and ionic bonds respectively.

D1   Type 1: Materials with low boiling points

In a gas or vapour the particles are in rapid and random motion – there cannot be any very strong links or attractive forces between them or they would stay close together. If a liquid readily becomes a vapour even around room temperature (in

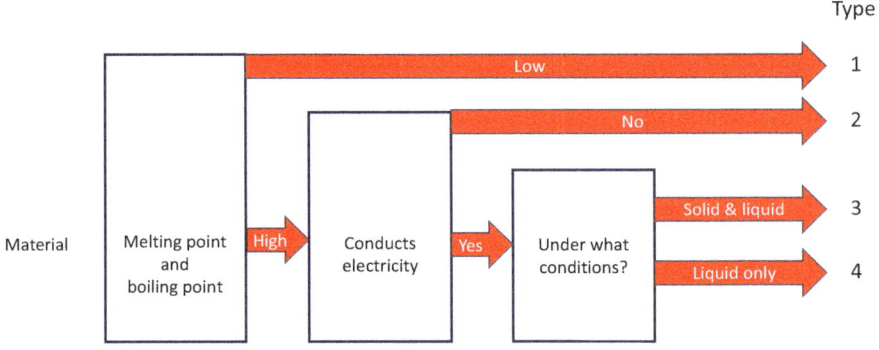

Figure 4.15 Flowchart to separate materials into the four types.

strong contrast to water, for example), then the links between the particles cannot be very strong in the liquid state either. If you have to heat a liquid strongly before it will become a vapour, then the particles must be held together fairly strongly.

Volatile compounds such as ethanol can be distilled. They evaporate and condense when they have been heated, but still they remain the same compounds. This means that although the links between the molecules are very weak, the links between the atoms within the molecules must be strong and not easily broken.

This condition is just what you would expect of the molecules of covalently bonded substances. Electronically, all the molecules are well satisfied, as the mechanism which holds the atoms together has left no tendency for further joining. The molecules are electrostatically neutral since they are composed of neutral atoms; the total number of electrons balances the total number of protons in any one particle. For a material to conduct electricity there must be movement of electrically charged particles, for example ions or electrons. As these covalent molecules are neutral, moving them about does not cause any movement of electric charge and so these substances are not conductors of electricity.

Light molecules move more easily than heavier ones. We would expect materials with low molecular mass to have lower boiling points than those with high molecular mass, because the heavy molecules will need more heat to get them moving fast enough to become a gas. This trend is shown in Table 4.3. (There are exceptions to it. Water is one exception.)

The fact that covalent substances can exist as liquids and solids as well as gases must mean that there are *some* forces of attraction between molecules. These are weak compared to the bonds between atoms in molecules and so are called **secondary bonds**. These forces often cause dirt to stick to objects (see the book in this series, *Cleaning*) and they account for the adhesion of glues (see the book in this series, *Adhesives and Coatings*). For now, it is enough to be aware that forces weaker than the **primary bonds** (forces of attraction *within* molecules) do exist and that they, too, cover a range of strengths. These forces hold polar molecules together. This is why substances with polar molecules such as water and ethyl alcohol, which are liquid at room temperature, need more heat put into them to make them boil than do substances of similar or greater molecular mass which have non-polar molecules (such as nitrogen, oxygen and carbon dioxide).

Table 4.3 Range of boiling points in low molecular weight and simple compounds.

Substance	Molecular mass	Boiling point (°C)
hydrogen	2	−253
nitrogen	28	−196
carbon dioxide	44	− 78
diethyl ether	74	+ 35
toluene	92	+111

D2   *Type 2: Non-volatile materials which do not conduct electricity*

It is unusual for a compound with molecules containing more than 40 atoms, or which has a molecular mass greater than 350, to be volatile. There is, however, no restriction on the size of covalent molecules, and natural and synthetic polymers such as protein and polythene may have molecular masses of many hundreds of thousands. Often such **large molecules** are made up by repeating relatively simple units in the form of a chain (polyvinyl acetate for example). There is then a range of possible molecular masses because the number of repeat units can vary from molecule to molecule. Molecular formulae such as $(CH_3COOCHCH_2)_n$ can be used to show that an uncertain but large number of units is repeated. A great many materials used in conservation are composed of giant covalent molecules; wool, silk, cotton, polyester thread, wood, leather, polyvinyl acetate (PVA), nylon, Perspex® (Plexiglas®), etc. The molecules of cellulose in cotton are composed of about 3000 repeated units each containing 21 atoms. Figure 4.16 shows part of its structure.

Materials with such big molecules cannot evaporate easily. The long chains may actually become physically tangled in these **long-chain molecules**. When the material is heated the primary bonds are eventually broken and chemical changes such as charring are frequently observed.

The properties of a substance composed of large covalent molecules will depend very much on their size and shape. If the molecules are relatively compact, they will stack together neatly and regularly in the solid and will form hard crystals (like sugar). If they are long, twisting, string-like molecules, they will tend to lie lengthways alongside one another but not necessarily in any orderly fashion. In this case, the solid may be very flexible and have strong directional properties. Examples are the fibres of wood and silk.

Diamond is a poor conductor of electricity and is not at all volatile. It can be burned in oxygen, though only at extremely high temperatures, to form carbon dioxide and nothing else. This expensive experiment shows that diamond is composed of nothing but carbon atoms and is chemically identical to the black carbon we are familiar with in black carbon inks. If the carbon atoms were not joined to one another in any way, we would expect it (with a molecular mass as low as 12) to be very volatile, even gaseous, at room temperature (compare the gas nitrogen

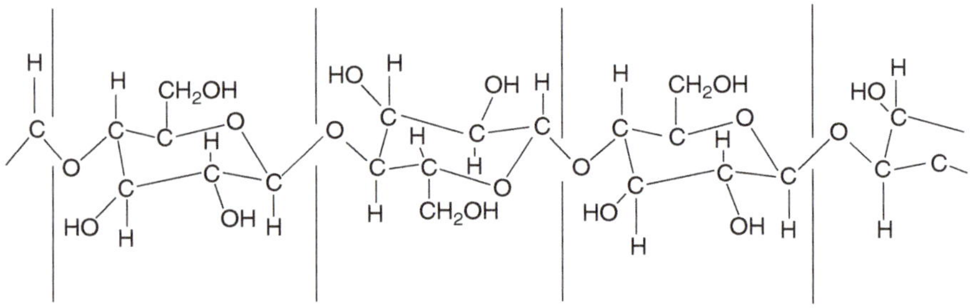

Figure 4.16 Part of the structural formula for cellulose $(C_6H_{10}O_5)_n$, showing how the units repeat.

which has a molecular mass of 28); but diamond is hard and crystalline with a melting point in excess of 3500°C. The molecule or crystal of diamond must be very big; what is its structure? If it contains covalent bonds the diamond molecule will contain carbon atoms with a valency of 4. There are several ways that large numbers of carbon atoms could be joined together so that each has four bonds attached to it. (Try to see how many you can find.) The rigidity and symmetry of diamond suggests that there are not long flexible strings of carbon atoms or large, flat, sheet-like molecules that could slide over one another. The actual structure of diamond in some ways resembles that of methane, where the C–H bonds point to the corners of an imaginary tetrahedron. The diamond 'crystal' is an infinite three-dimensional **lattice** of carbon atoms each joined by four tetrahedrally arranged bonds to four other carbon atoms, which means that 'molecule' is not such a good word for it. 'Giant molecule' or 'extended covalent lattice' would be better descriptions.

This structure exists throughout a single piece of the solid and so it does not form a molecule in the sense discussed so far. For this reason, the word *molecule* is seldom used for such giant three-dimensional arrays. Thermosetting polymers like casting resins such as styrene-based polyester, or Bakelite®, form similar continuous three-dimensional covalently bonded structures.

The mineral silica is another example of these extended covalent lattices. It is composed of atoms of the element silicon (some Europeans whose first language is not English call this element, Si, silicium) combined with twice the number of oxygen atoms. Since the elements are in the ratio 1:2 it is given the formula SiO_2, but this does not imply that there are individual SiO_2 molecules. Silica can have several structures, quartz (sand) being one of them. In all the structures the silicon–oxygen bonds are tetrahedrally arranged and there is an oxygen atom between any two silicon atoms. (Silicon has a valency of four and oxygen a valency of two.) These materials are non-conductors of electricity, that is, insulators. This is because the particles are not free to move in the rigid bonding framework. The electrons are all involved in bonding and the atoms are neutral. (Terminology can be confusing; note that this paragraph has not been discussing *silicone*, which is an organic material, encountered in conservation workspaces as silicone rubber used for taking a mould of a textured surface to replicate it, and in laboratories as silicone grease.)

D3   *Type 3: Materials with high melting points which conduct electricity in the solid state*

Materials with high melting temperatures which conduct electricity in the solid state must have strong bonds holding the atoms together, but there must also be some freely mobile electrically charged particles. Most such materials are metals: the previous description of the metallic bond fulfils these requirements. The atoms are held tightly together, while the free (or valence or de-localised) electrons which become the common property of all the atoms can readily move within the metal under the action of any electric driving force (meaning, an applied voltage). Other properties of metals, such as their lustrous appearance, good conductivity of heat, ductility and malleability can also be explained in terms of a gas or sea of free electrons, as noted earlier. Even simple spherical atoms can be arranged regularly in three dimensions in several different ways when they are combined together as a compound, because the 'diameter' of atoms increases with the number of protons and neutrons in the atom. The atoms in ductile metals are aligned in planes. There is no strong bonding directly between any two atoms, so large groups of atoms can move relative to one another along these planes without breaking any specific bonds. When this happens, the crystals become deformed.

Graphite, one of the forms in which carbon is found, has a structure which has some similarity to that of the metals. The carbon atoms are joined by covalent bonds into planar networks of hexagonal rings, as shown in Figure 4.17.

This representation would seem to imply that carbon atoms have a valency of only 3, not 4 as discussed earlier, but the orbitals of each atom are able to overlap in such a way that the fourth outer electron is not localised in a covalent bond but can travel throughout the sheet of atoms as a free electron. Like the metals, graphite conducts electricity by the movement of electrons. The sheets of carbon atoms are stacked one on top of another, but there are only weak secondary bonds between the sheets. The sheets readily slide over one another, which is why graphite is so soft. This structure of sliding sheets of atoms explains why graphite is used as a lubricant and why a graphite pencil leaves a mark on paper (so-called 'lead pencils' contain graphite, not lead; although both metallic lead and silver have historically been used for drawing too).

D4   Type 4: Substances with high melting temperatures which conduct electricity only in liquid states

The ionic bond can give these properties – high melting temperatures and electrical conduction in liquid states only. High melting temperatures imply strong forces

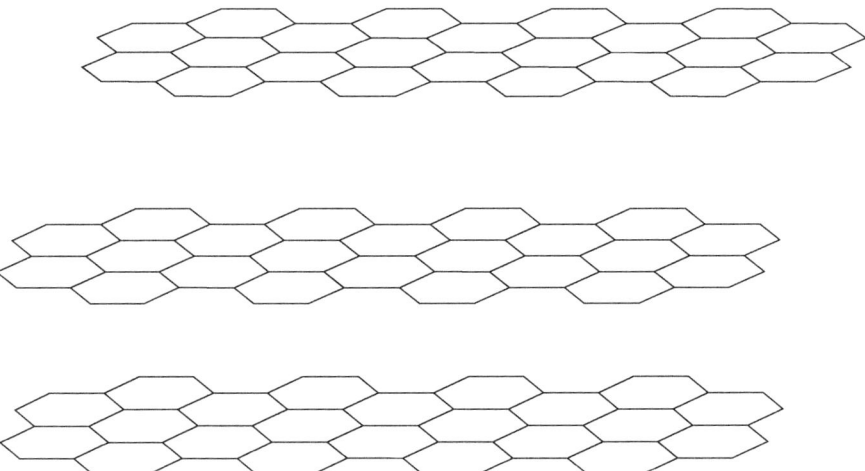

Figure 4.17 The structure of graphite.

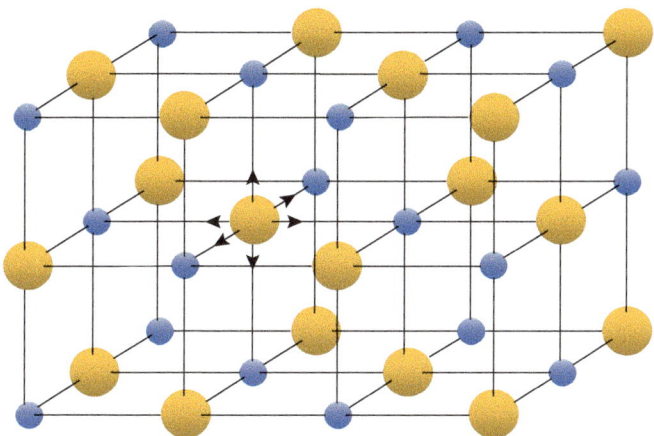

Figure 4.18 The structure of sodium chloride. Each negative ion (large) is surrounded by six positive ions (small) as indicated by the arrows, and *vice versa*.

between all particles. The electrostatic attraction between +ve and −ve ions holds them together to form solid crystals. For electrical conduction the ions themselves must be able to move, there being no free electrons as in metals. Ions can only move if the strong bonding of the crystals is disrupted by melting or by dissolving into water.

The structure of ionic crystals reflects an orderly arrangement of ions controlled primarily by their charges, sizes and shapes. Common salt, for example, contains sodium ions (Na^+) and chloride ions (Cl^-) in equal numbers. They have to be equal to maintain overall charge neutrality: sodium atoms have one electron to lose and chlorine atoms need one to form a full shell. Figure 4.18 shows the arrangement of ions in the crystal. As you can see, the cations and anions alternate throughout the structure.

The section of crystal illustrated in Figure 4.18 lets you see that each Cl^- ion is surrounded by six Na^+ ions. This is the basis of the strong attractive force. The grid of straight lines is only there to show the geometrical arrangement; there are no directional bonds. The electron clouds of the ions would in fact be touching.

You will notice that Cl^- ions are depicted as larger than Na^+ ions. It is generally true that cations are small because when electrons have been lost the nucleus holds the remaining cloud a little more tightly than before. Conversely, the extra electrons pushed into anions swell the cloud a little. The larger size of anions, their mutual repulsion and the non-spherical shapes of those which are not single atoms, largely determine the particular arrangements in a crystal. (Lead white $Pb_2(OH)_2CO_3$ is an example where the anions are of two different kinds, and the arrangement of anions and cations in that pigment is rather complicated.)

5   Relating chemical names to structure

Your newly acquired knowledge of bonding can immediately be put to work in understanding the rationale behind systematic chemical names. The major sub-classification is between inorganic materials and organic ones and this chapter aims to show the basis of naming for a reasonable range of each. It cannot be complete but will show you which names for materials have meaning in terms of defining chemical composition. The chapter includes older naming systems for chemicals. This is not a recommendation to use older names or trivial names, but it might serve as a useful reference and an aid to understanding older treatment reports, and even more so, historic literature on recipes for manufacturing processes for many materials ranging from pigments and dyes to industrial products.

A   Forming chemical names

In everyday language we use slightly different words to carry shades of meaning or to play different grammatical roles in a sentence. (*Conserve, conservator, conserva-tive, conservation.*) The exact construction of each word follows recognised conventions but is really quite arbitrary. *Because* of the conventions we can recognise differences in meaning and function. To understand **chemical names** you have to become similarly sensitive to small differences between words and be able to interpret them. Thus *sulfuric, sulfurous, sulfide, sulfite* and *sulfate* all obviously have something to do with the element *sulfur*, but each means something slightly different, yet specific. The problem is that the information is coded and, because the code has been arbitrarily chosen, you just have to learn it in English. In many European languages, the spelling for these words is pretty similar when they are translated, so a chemical suppliers' catalogue in another language is pretty easy to read.

The main objective behind chemical naming is to describe molecular structure. To do this, new names have been created to describe groups of atoms which, although not molecules in their own right, confer recognisable reaction characteristics on a molecule and often remain intact during a chemical reaction. These groups are known as **functional groups**. In an exercise you were asked to draw a structure for the organic compound methylamine – the name tells those who know the code that the molecule consists of a methyl group, CH_3-, bonded to an amine group, $-NH_2$. It must be CH_3-NH_2: the name indicates the structure.

DOI: 10.4324/9781003261865-5

Organic chemistry is defined as the chemistry of carbon and its compounds. It deals predominantly with covalent molecules made from a very small range of elements (carbon, hydrogen, oxygen and nitrogen being the most common). Solvents, polymers and therefore plastics are all organic materials that are mostly made from the first three of these elements, which is a reason for the traditional saying in chemistry, that 'like dissolves like'.

Inorganic chemistry is the study of the reactions of all elements other than carbon. A great many inorganic compounds are ionic and so the anions and cations are the groups used as the basis for the naming systems. Although these names make up quite an extensive vocabulary only a limited selection will be used in this chapter, to show you how small variations in names can imply differences in structure. Often you will find fragments of element names and group names joined with prefixes and suffixes to form new group names.

The systems for inorganic and organic naming will be dealt with separately, although they do have something in common which is the use of **suffixes** and **prefixes** derived from both Greek and Latin. Most common are the Greek-derived prefixes telling how many atoms or groups there are in a molecule:

---

mono one
di two
tri three
tetra four
penta five
hexa six
hepta seven
octa eight
nona nine
deca ten
poly many, in fact far more than 10 or 100 or 1,000.

The compound names *carbon monoxide, carbon dioxide, monosodium glutamate*, give some examples of their use.

B Inorganic compounds

Some simple **covalent molecules** have trivial names which must be learned, as they give no real clue to the elements present or to the way in which they are combined. Water, H_2O; ammonia, NH_3; and ozone, O_3 are examples. Systematic names such as *sulfur dioxide* SO_2 are self-explanatory. More complex inorganic covalent molecules are named using systems related to those described for the organic compounds.

Calcium hydroxide, lead carbonate and copper sulfate are names which may be familiar. Notice that each is a **two-word name**, the first word being the name of a metal. Because you now know that metals form **ionic compounds** with the metal atoms becoming positive ions, you might correctly deduce that the other part of each name describes a negative ion. Let us look at the way anion names are formed.

B1   The ending -ide

Two negative ions for which you have already seen the electronic structures are oxide and chloride. These are the ions formed when an atom of oxygen or of chlorine acquires enough electrons to fill its outer shell. Oxygen needs two electrons to form O^{2-} and chlorine one electron to form Cl^-. These are examples of negative ions formed from a single atom of an element and there are several others:

fluoride F^-
chloride Cl^-
bromide Br^-
iodide I^-
hydride H^-
oxide O^{2-}
sulfide S^{2-}
nitride N^{3-}

There are a few anions whose names end in *-ide* which contain atoms of more than one element. An example of such a stable negative ion group is hydroxide, OH^-.

B2   The endings -ite and -ate

When the ending *-ite* does not occur in a two-part name but is found in a single word name (such as the minerals *malachite* or *azurite* which are both traditional pigments that have been used in many cultures), it shows that this is the trivial name for a particular mineral. (Occasionally it occurs in commercial and often trademarked names such as *Dynamite®*, which are correctly give a capital letter because of their trademark.)

In compound names the endings *-ite* and *-ate* are commonly found in conjunction with part of an element name. There are examples where both endings occur such as *nitrate* and *nitrite*, *chlorate* and *chlorite*. There are others such as *carbonate* and *silicate* which only occur as *-ates*. All these are names of negative ions formed by oxygen combining with the indicated element:

carbonate CO_3^{2-}
phosphate PO_4^{3-}
silicate SiO_4^{4-}

The endings *-ite* and *-ate* distinguish between ions containing different amounts of oxygen. An *-ite* ion always contains one oxygen atom fewer than the corresponding *-ate* ion, and so for the examples mentioned, the formulae for the ions are:

Table 5.1 Some names for *-ite* and *-ate* ions.

-ite	-ate
nitr*ite* NO_2^-	nitr*ate* NO_3^-
sulf*ite* SO_3^{2-}	sulf*ate* SO_4^{2-}
chlor*ite* ClO_2^-	chlor*ate* ClO_3^-

Further modification by prefixes

This simple rule is not sufficient as some families of ions have more than two members. Prefixes were historically used to increase the number of names. They have the following meanings, although the IUPAC name is far more common today than using these prefixes.

Table 5.2 Historic prefixes used for naming ions.

hypo-	containing *less* oxygen than ...
per-	containing *more* oxygen than ...
thio-	containing a sulfur atom in place of an oxygen
bi-	containing twice as many anions
sesqui-	containing 1.5 times as many anions

There are four ions formed by combinations of chlorine and oxygen: ClO^-, ClO_2^-, ClO_3^-, and ClO_4^-. The middle of these are already named above, but names are needed for the ion ClO^- containing *less* oxygen than chlorite and for ClO_4^- which contains *more* oxygen than chlorate. Using the relevant historic prefixes, the whole family of names is therefore:

ClO^- hypochlorite
ClO_2^- chlorite
ClO_3^- chlorate
ClO_4^- perchlorate

The prefix *per-* is also met in the peroxide ion, O_2^{2-}, which contains more oxygen than the oxide O^{2-}. (Hydrogen peroxide, H_2O_2, contains more oxygen than water, H_2O.) *Per-* is also found in permanganate, MnO_4^-, as in potassium permanganate, $KMnO_4$, a chemical little used today in conservation. Since the *per-* compounds such as hydrogen peroxide contain excess oxygen, it is not surprising that they are strong oxidizing agents, which is precisely why they are rarely used today to treat objects.

In the thiosulfate anion, one oxygen atom in the sulfate ion, SO_4^{2-}, has been replaced by a sulfur atom to form $S_2O_3^{2-}$. The salt $NaHSO_3$ is now called *sodium hydrogen sulfite* in preference to *sodium bisulfite*.

B3 Words ending in -ous and -ic

The suffixes *-ous* and *-ic* are found in the first word of two-part names. They usually refer to types of cation and distinguish different electronic states of a particular metal. The words are not always formed directly from the element name. Many of the metals that have been in common use since antiquity do not have names ending in -um or -ium, whereas many of the other metallic elements, named in the late eighteenth century or more recently when they were first recognised as distinct metals, do. Where the common name is difficult to use in this way, the *-ous* and *-ic* words are formed from the Latin name for the metal.

Table 5.3 Symbols for metals, derived from their Latin names as once used by scholars and alchemists.

Metal	Latin name	Symbol
copper	*cuprum*	Cu
gold	*aurum*	Au
iron	*ferrum*	Fe
lead	*plumbum*	Pb
silver	*argentum*	Ag
tin	*stannum*	Sn

Table 5.4 Ions with multiple valencies.

Positive				Negative	
Cu^+	1 valent	Cu^{2+}	2 valent	SO_4^{2-}	2 valent
Fe^{2+}	2 valent	Fe^{3+}	3 valent	O^{2-}	2 valent
Sn^{2+}	2 valent	Sn^{4+}	4 valent	Cl^-	1 valent
Pb^{2+}	2 valent	Pb^{4+}	4 valent	CO_3^{2-}	2 valent

An example you may well have met concerns the corrosion of copper. Copper forms two oxides, a dull red substance, Cu_2O, and a black substance, CuO. Since oxygen needs two electrons to achieve a full electron shell you can see that in Cu_2O each copper atom is giving one electron while the one copper atom in CuO gives two electrons. The names of these compounds are *cuprous oxide* for Cu_2O (or copper(I) oxide), and *cupric oxide* for CuO (or copper(II) oxide). The suffix to use in the lower valency state is *-ous* and in the higher valency state, *-ic*. In the *-ic* condition an atom has given away more of its electrons than in the *-ous* state.

You also meet *-ous* and *-ic* words when describing acids. You can regard **acids** as combinations of hydrogen with negative ions. The reason for regarding them as special is that when the acid is pure the negative part is forced to make a *covalent* bond with the hydrogen. Thus H_2SO_4, sulfuric acid, is a true molecule while $CuSO_4$ is a part of a crystal. Another name is needed for the molecule H_2SO_3 to distinguish it from sulfur*ic* acid; it is sulfur*ous* acid.

B4   Salts

The combinations of metal ions and negative ions from acids are collectively known as *salts*. 'Common salt', sodium chloride, is just one example. It is instructive to visualise how it could be formed in a reaction between hydrochloric acid and sodium hydroxide. (Note that this is a thought experiment: both of these chemicals require a risk assessment, and the use of personal protective equipment and a fume hood before handling them separately, and even more so before combining them at full strength. Any experiments at mixing these solutions are better done with 0.1M or even weaker solutions, but still using protective equipment.) When in solution in water, hydrochloric acid, HCl, breaks down to form the ions H^+ and

Cl^- while sodium hydroxide, NaOH, dissolves to give Na^+ and OH^- ions, producing a mixture of Na^+, Cl^-, H^+ and OH^-. Almost all the H^+ and OH^- ions join up to form *covalently* bonded water molecules, H_2O, leaving Na^+ and Cl^- only. If the solution dries out these ions cohere electrostatically to form crystals of salt.

Acids whose name is of the form *hydro ... ic* react to give salts which end in *-ide*; hydrobromic acid produces bromides. Acids whose names are formed simply from the element name plus *-ic* give rise to *-ate* salts; phosphoric acid produces phosphates. Acids ending in *-ous* produce *-ite* salts; sulfurous acid gives sulfites. These are simply examples of the naming system, not recommended conservation materials.

The cationic part of a salt is not always a metal. Ammonia, NH_3, reacts with acids to form salts containing the cation NH_4^+. This is called the *ammonium* ion, the *-ium* ending showing that it forms salts like a metal, such as sodium.

C   Organic compounds

Organic compounds may contain large numbers of atoms of only a few elements. There are 35 possible structures for the molecule C_9H_{20}, which contains 29 atoms though only two different elements. Obviously, a naming system which uses only the element name as a root with a few prefixes and suffixes is not sufficient. Several ways of overcoming this difficulty have been attempted.

Before molecular structure was understood, compounds were often given a **common** or **trivial name** derived from their source; acetic acid was made from vinegar, and its common name is formed from the Latin word for 'vinegar'. These names do not indicate structure. When a new compound was made by a reaction of, say, acetic acid, the new product would be given the prefix *acet-* and an ending to show what sort of change had taken place. This introduces the idea of a group or part of a molecule which remains intact throughout the reaction. Other groups which remain unchanged in a reaction are those derived from hydrocarbons, like *ethyl* and *methyl* from ethane and methane. A naming system has been devised which is based on these hydrocarbon names. The IUPAC system works by looking at all carbon-containing (that is, organic) compounds as a backbone of carbon atoms with various functional groups attached to it. By using numbers you can designate which carbon atom each functional group is attached to.

Although they are very precise, the newer naming systems can be very cumbersome and have not been universally adopted, especially in speech. The result is that for common substances trivial, old system and new system names are quite often used side by side. Thus you find that acetone, dimethyl ketone and propan-2-one are the same compound, and any of these terms might be found in conservation publications from different decades, while many conservators only ever call this solvent acetone when talking. Consequently we have to discuss all three systems in this book.

C1   *Names derived from those of hydrocarbons*

The simplest organic molecules are the **hydrocarbons**, made of carbon and hydrogen alone. They are most commonly met as fossil fuels (petrol, diesel oil, natural gas) which are mixtures of several different molecules, and they are used as the source for making petrochemicals. One (largely) hydrocarbon mixture commonly used in several areas within conservation is known, and variously marketed in different countries, as white spirit, white spirits, Stoddard solvent, solvent naphtha, and in the context of hardware stores, as turpentine substitute.

You need to learn the names and structures of some simple hydrocarbons, as the names of functional groups containing only C and H atoms such as *methyl* and *ethyl* are derived from them. You are already familiar with methane, CH_4. This is the smallest hydrocarbon molecule and has just one carbon atom. The names and structures of some other hydrocarbons are given below: this series of hydrocarbons is known as the **alkane series** or simply as **alkanes**.

When more than four carbon atoms are joined in a line the names again correspond to Greek numbers:

Table 5.5 The first four members of the alkane series.

Number of carbon atoms	Formula	Name	Structure
1	CH_4	methane	
2	C_2H_6	ethane	
3	C_2H_8	propane	
4	C_4H_{10}	butane	

five carbons gives pentane: C_5H_{12}
six carbons gives hexane: C_6H_{14}
seven carbons gives heptane: C_7H_{16}
eight carbons gives octane: C_8H_{18}
nine carbons gives nonane: C_9H_{20}

If one of the hydrogen atoms comes off the end of such a hydrocarbon molecule, a functional group which can join on to something else will be formed. The free bond is denoted by an unattached dash as shown in Figure 5.1.

Table 5.6 The first five members of the alkyl group.

Number of carbon atoms	Name	Formula
1	methyl	CH_3-
2	ethyl	C_2H_5-
3	propyl	C_3H_7-
4	butyl	C_4H_9-
5	pentyl (the pentyl group was formerly known as *amyl*)	$C_5H_{11}-$

As a class these groups are called the **alkyls**. If, in a formula, an unspecified alkyl group is to be shown, it is usually represented by the letter R–.

In the parent hydrocarbons, as the molecular chain gets longer, different physical properties develop, as shown in Table 5.7.

This trend is followed to some extent by compounds containing the groups with the same numbers of carbon atoms. So the name begins to tell you something about the physical properties of the compound.

The group to which the alkyl group R– is attached is what dictates the chemical behaviour of the compound. The reactive groups of atoms are called **functional groups**. An example is the –OH group. This is the functional group in the class of compounds R–OH which are known as the **alcohols**. The name of a specific compound is formed by adding the name of the alkyl group to the class name. Thus C_2H_5- (ethyl) joined to $-OH$ (alcohol) makes C_2H_5OH, ethyl alcohol, which is more correctly called ethanol.

Table 5.7 Properties and uses of hydrocarbons.

Number of carbon atoms	Physical state	Name	Uses
1 to 4	gas	natural gas	gas fuels
7 to 9	volatile, mobile liquid	petrol (gasoline)	automotive fuel
10 to 12	less volatile, less mobile liquid	paraffin (kerosene)	aircraft fuel
13 to 18	sluggish liquid	diesel oil	heavy engine fuel
20s	slimy liquid	oil	lubricating oils
several tens	soft solid	wax	candles
several hundreds	stiffer solid	Polythene (more correctly called polyethylene)	thermoplastics

In each group, such as the alcohols, the chain of carbon atoms forms a kind of 'tail' on the molecule, and the longer it gets, the more it influences the chemical properties of the whole molecule: the head has less effect and the tail more. Compound series derived from the alkanes, for example, are also known as **aliphatic hydrocarbons**, where the word aliphatic refers to the long straight tail of carbon atoms that has two hydrogen atoms attached on either side.

Exercise

1. Give systematic names to these compounds:
    (a) CH_3OH
    (b) C_3H_7OH
    (c) $C_2H_5NH_2$
    (d) CH_3OCH_3

Other functional groups

Propane, C_3H_8, has two different 'types' of hydrogen atom, those on the end carbon atoms and those on the central one. If a hydrogen atom from one of the end carbons is removed by chemical reaction the *normal propyl* functional group, written as '**n- propyl**' is formed, $CH_3CH_2CH_2-$. If a hydrogen atom from the central atom is removed the **isopropyl** functional group results (Figure 5.1).

Isopropyl alcohol, sometimes known as IPA, more correctly called isopropanol, and to be completely correct called propan-2-ol (as will be discussed below), which indicates the position of the carbon that has lost the hydrogen atom, is a common solvent used in conservation treatments.

Another hydrocarbon-derived functional group whose name you will be familiar with is the **vinyl** group (Figure 5.2). This is derived from ethylene. Combined with –OH it becomes vinyl alcohol and when combined with –Cl, vinyl chloride. These names are used with the prefix *poly-*, which means many, to describe the compounds polyvinyl alcohol (abbreviated sometimes to either PVAL and PVOH) and

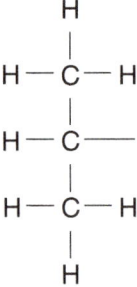

Figure 5.1 Isopropyl group.

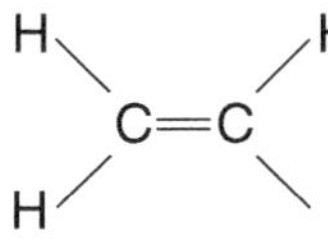

Figure 5.2 Vinyl group.

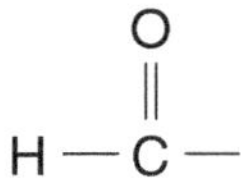

Figure 5.3 Form- functional group.

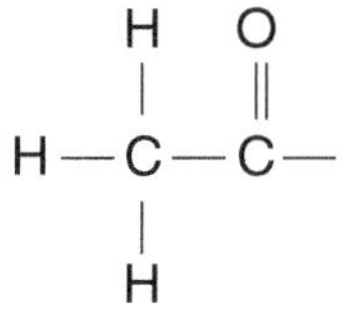

Figure 5.4 Acet- functional group.

polyvinyl chloride (PVC), the first used as a conservation adhesive and the second as plastic sheeting often made into pockets or bags. **Polymerization**, the joining together of units or functional groups such as vinyl alcohol into long molecular chains, will be discussed in full in a later book in the series, titled *Adhesives and Coatings*.

C2   The acet- and form- names

Acetone and formaldehyde are organic compounds you will know, the latter being a volatile compound released from wood composite products, which can increase the corrosion rate of metals. Formic acid was once used to clean corroded silver, while acetic acid was used to 'fix' fugitive dyes before washing textiles, to make it less likely that the dye would wash out during cleaning with a water-based solution. Their names are examples of the class of compounds beginning with *form-* and *acet-*. All the *form-* compounds contain the group shown in Figure 5.3.

All the *acet-* compounds contain the group shown in Figure 5.4.

The name of the compound depends on what other atom or functional group is attached to the free bond. The names and structures of some of these compounds are given in Table 5.8.

While *acet-* and *form-* are not systematic names, they are the terms used most often in conservation for these particular compounds. These groups of names use a stem which comes from the trivial name for the organic acids. Some other acids give rise to compounds whose names are formed in the same way and whose names

Table 5.8 Form- and acet- compounds encountered in conservation treatment, or involved in deterioration of some materials found in collections.

Group added	Chemical class	form- Names	form- Structure	acet- Names	acet- Structure
H	aldehyde	formaldehyde		acetaldehyde	
OH	organic acid	formic acid (methanoic acid)		acetic acid (ethanoic acid)	
O-C$_2$H$_5$	ester	ethyl formate		ethyl acetate	
O-Na	salt	sodium formate		sodium acetate	
CH$_3$	ketone	-	-	acetone	

may be familiar. The class of compounds called aldehydes are all named in this fashion, as shown in Figure 5.5.

The acids, whose names all end in -*ic*, can form salts whose names all end in -*ate*. **Organic acids** also react with alcohols to form a class of compounds called **esters**. You might like to imagine these as the organic equivalent of salts as there is some similarity between the reaction shown below and that described in section B4. They are not bound as ionic compounds but by covalent bonds: Figure 5.6 gives an example.

Palmitic acid $C_{15}H_{31}COOH$ and stearic acid $C_{17}H_{35}COOH$ form palmitates and stearates which are found in domestic soap as sodium salts, and in ageing paint films made from drying oil and lead white as lead soaps. These reactions are usually fairly slow in the case of oil paintings, taking place over decades or centuries before they affect the paint properties dramatically. The formation of an oil paint film can be regarded as ester formation in one sense – however, the molecules involved here are larger, and other terms can equally correctly be used to describe the chemical reaction. Outside the conservation field, they are important for developing flavours in foods. They account to some extent for the taste of wine as it matures, the esters giving complex flavours as they replace the acid present in young wines.

Vinyl acetate is the basic unit of the familiar polyvinyl acetate (PVA) group of adhesives (Figure 5.7).

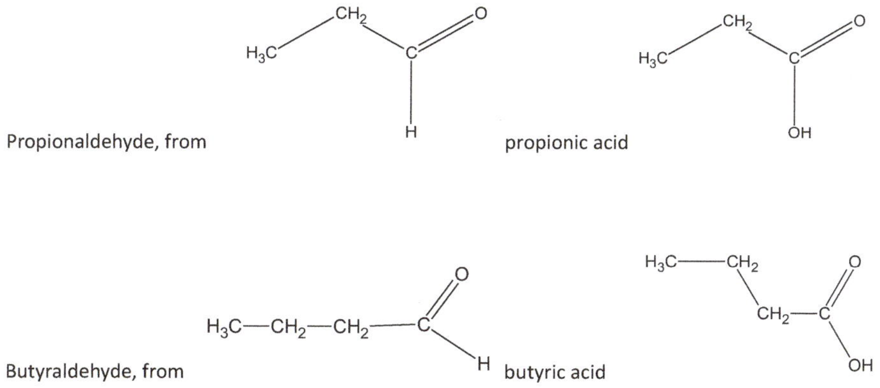

Propionaldehyde, from   propionic acid

Butyraldehyde, from   butyric acid

Figure 5.5 Propion- and *butyl*- aldehydes.

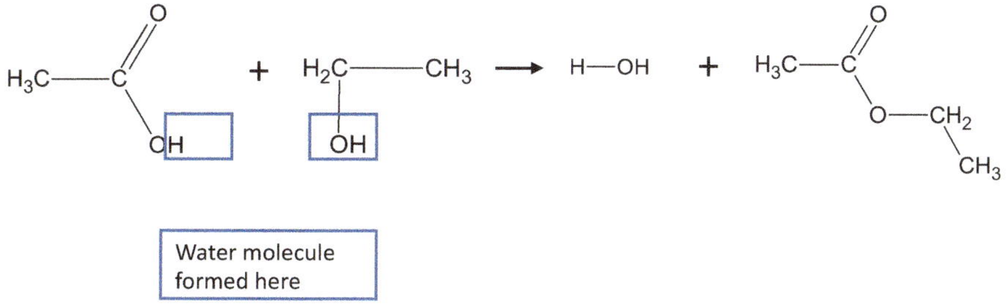

Water molecule
formed here

Figure 5.6 Acetic acid plus ethyl alcohol forms ethyl acetate plus water.

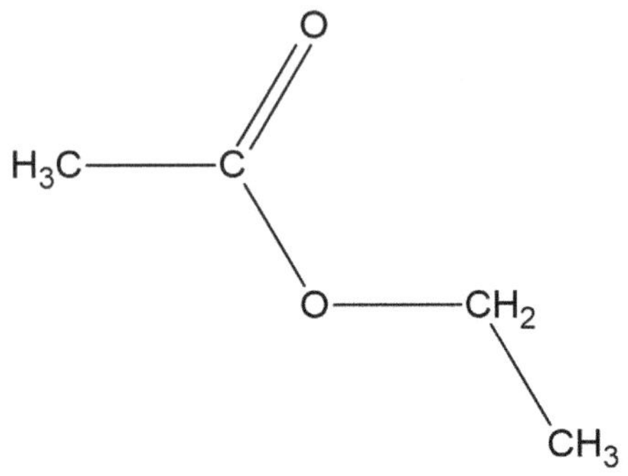

Figure 5.7 Structure of vinyl acetate.

C3   *The modern IUPAC naming system*

The first attempts at a system of naming which indicated molecular structure failed largely because too many starting points were used. We have looked at two of them – the names based on hydrocarbon names and those using the early names like *acet-* for organic acids. With hindsight, another reason is that organic chemists soon made and analysed the structure of so many complicated molecules that a system built on pairs of functional groups simply could not cope.

The key to the new system is to count the number of carbon atoms linked together in a chain. The appropriate fragment of the hydrocarbon name then forms a basis for the compound name depending on what atoms or functional groups are attached. A code is used to describe groups substituted for hydrogen (Table 5.9). You will notice some connections with the earlier names.

Thus there are those regarded as derived from methane (shown in Figure 5.8), which are shown in Table 5.10.

Those derived from ethane have two carbon atoms, as shown in Table 5.11.

Table 5.9 Functional groups substituted for hydrogen. (The carbon atoms in the last three are counted as part of the hydrocarbon fragment and included in the count that gives its name.)

$-OH$	gives the ending *-ol*
$-CHO$	gives the ending *-al*
$-C = O$	gives the ending *-one*
$-COOH$	gives the ending *-oic acid*

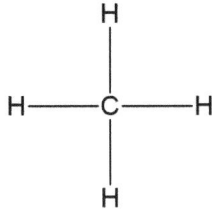

Figure 5.8 Structure of methane.

Table 5.10 Naming of compounds derived from methane.

New systematic names	Molecular structure	Old systematic names
methanol		methyl alcohol
methanal		formaldehyde
methanoic acid		formic acid

Table 5.11 Naming of compounds derived from ethane.

New systematic names	Molecular structure	Old systematic names
ethane		-
ethanol		ethyl alcohol
ethanal		acetaldehyde
ethanoic acid		acetic acid

Three carbon atom chain molecules have their names based on propane $(CH_3CH_2CH_3)$, as shown in Table 5.12.

The numbers indicate the carbon atom to which the functional group is attached. So, in propan-2-ol, the OH is attached to the second C as you read the molecular formula from left to right. Here you can begin to see the strength of the system in writing, but also its clumsiness in speech.

In some compounds the carbon chain is interrupted by an atom of a different element, as in ethers and esters. The substance with the old systematic name of ethyl acetate has the structure shown in Figure 5.9. It is an ester of ethanoic acid and is now called ethyl ethanoate.

Ethers are regarded as substituted hydrocarbons. Thus diethyl ether (its trivial name is just 'ether') is thought of as ethane with one H replaced by an ethoxy (C_2H_5-O-) group, and so $C_2H_5-O-C_2H_5$ becomes ethoxyethane shown in Figure 5.10.

This new system of naming will not be pursued further here, although it is hoped that you can now see that it contains rules which are well enough defined to express very complicated structures – and this is precisely its function. Except in simple cases (such as ethanol) the new systematic names do, if anything, encourage the continued use of trivial names in casual scientific as well as conservation conversation.

Table 5.12 Naming of compounds derived from propane.

New name	Compound	Old name
propan-1-ol	$CH_3CH_2CH_2OH$	n-propyl alcohol
propan-2-ol	$CH_3CH(OH)CH_3$	isopropyl alcohol
propanal	CH_3CH_2CHO	propionaldehyde
propan-2-one	CH_3COCH_3	acetone

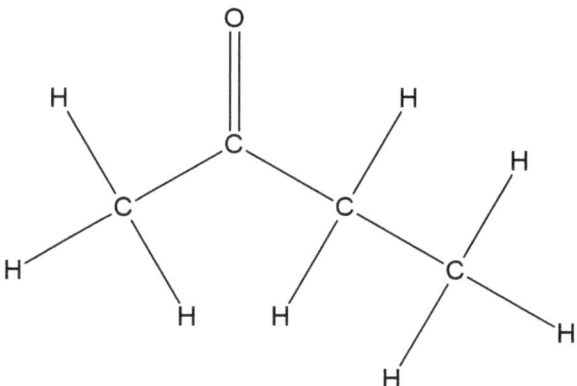

Figure 5.9 Structure of ethyl ethanoate.

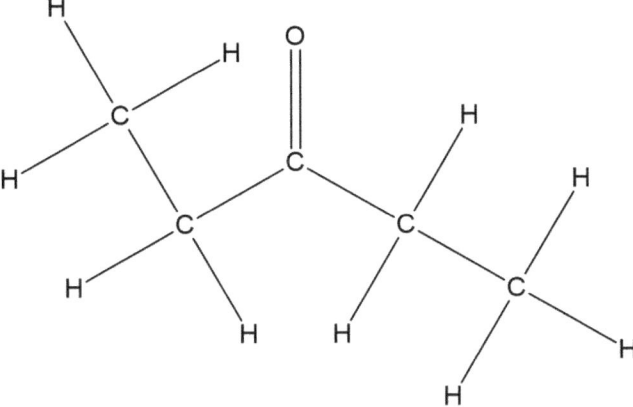

Figure 5.10 Structure of ethoxyethane.

C4 Benzene derivatives

All the organic compounds described so far have been formed from straight chains of carbon atoms. There is also an important group of chemicals in which the carbon atoms are joined together in a ring.

The name-ending *-ane* is used for hydrocarbons which contain only single bonds (methane, butane). The suffix *-ene* refers to hydrocarbons with double bonds in their structure (such as ethylene, now correctly called ethene). The ending *-ene* also occurs in the hydrocarbons benzene, toluene and xylene which have all been much used in the past, despite health and safety concerns, for removing coatings and varnishes, and as solvents for dissolving polymers for coatings and varnishes. (Xylene is still used in many countries for dissolving some polymers that form coatings and adhesives, because it is the least harmful of these three.) Benzene is the most toxic of the three and is banned in conservation workplaces and in most industries too, but its structure is key to understanding an important concept about chemical bonding, and thus it is introduced here.

Benzene, C_6H_6, has a molecule with a **ring** of six carbon atoms and originally, to fit with the idea of carbon having a valency of 4, a structure with alternate single and double bonds was proposed and was called a 'benzene ring'. You can see that the double bonds in the ring could justify the *-ene* ending on the name. However, this compound does not show the chemical reactions typical of double bonds. With the advent of ideas concerning molecular orbitals came a realisation that the carbon ring is made of single covalent bonds with the remaining six electrons de-localised and moving freely, so that all are equally shared by the carbon atoms, forming a resonance structure. Such rings are very stable and occur in a host of compounds (collectively known as **aromatic compounds**). As it is such a common structural element, the symbol shown in Figure 5.11 is now frequently used. The hexagon indicates the single bonds and the circle denotes the six free electrons. Notice that even the atom symbols have gone; a C – H is assumed at each corner of the hexagon.

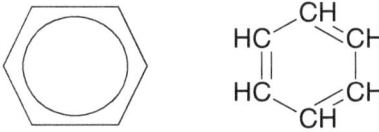

Figure 5.11 (a) The modern symbol to represent the structure of benzene. (b) Another representation of its bonds.

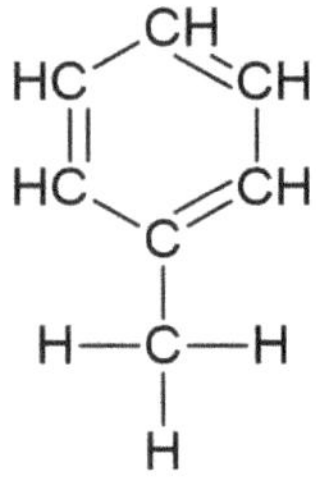

Figure 5.12 Structure of toluene.

Toluene C_7H_8 can be thought of as derived from benzene by the substitution of a methyl group (CH_3-) for one of the hydrogen atoms, so its structure is as shown in Figure 5.12.

Toluene is thus $C_6H_5 - CH_3$, since one of the hydrogens in Figure 5.11 has been substituted with a methyl group. The functional group $C_6H_5 -$ is called the **phenyl** group, so toluene could correctly be called either phenyl methane or methyl benzene.

Xylene has two such substitutions and can therefore exist as three isomers, because three distinct arrangements are possible for the methyl groups. Commercial xylene is a mixture of all three types, which are shown in Figure 5.13. The older names for the three types were *ortho*-xylene, *meta*-xylene, and *para*-xylene. In the IUPAC system the carbon atoms in the ring are numbered. The three possible isomers of xylene become 1,2,dimethyl benzene, 1,3,dimethyl benzene and 1,4,dimethyl benzene. (It is not important which C atom is called 1 in this context. Whichever one it is, the others are counted round from it by the shorter way.)

Phenyl amine or amino benzene is known as aniline, which is shown in Figure 5.14, a compound very important as a precursor molecule in nineteenth-century dye manufacture, although the term 'aniline dye' has been equally notorious since that time as a shorthand name for a class of colourants with poor stability when exposed to light. In fact, an enormous number of natural dyes are aromatic compounds that include several six-carbon rings in their structure, each ring joined to another by either one or two of the six bonds, which leaves the other bonds available for different functional groups to attach themselves to.

The group of solvents including benzene, xylene, toluene and other heavier ones with two benzene rings is known collectively as **aromatic solvents**. It is a common feature of aromatic compounds that they should be regarded as potentially

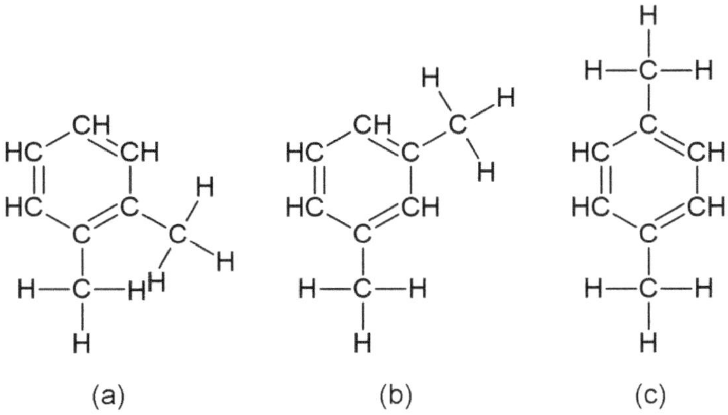

Figure 5.13 Structures for xylene: (a) 1,2,dimethyl benzene, (b) 1,3,dimethyl benzene, (c) 1,4,dimethyl benzene.

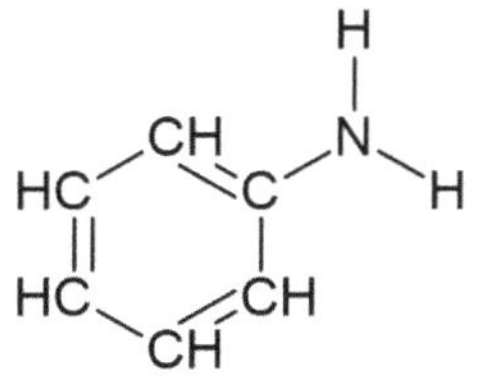

Figure 5.14 Structure of aniline.

hazardous to human health, because some of them are suspected of causing cancer or harm to the human foetus, the collective name for such chemicals being **carcinogens** in the first case, and **teratogens** in the second. Benzene certainly does cause cancer – and this became clear when it was industrially important in the manufacture of dyes. It has long been avoided by conservators. Toluene is very much suspected of being carcinogenic and it should be substituted with another solvent. Xylene does not have the all-clear in this respect either, which is why its use has declined so much in recent decades: it should not be used if it can be easily substituted with another solvent or mixture. Many of the well-known solvent blends sold in Europe and the USA, with trade names such as white spirit or the Shellsol® series (which comes in different grades, with some grades no longer made and new ones introduced from time to time) include a proportion, or more accurately a specified upper and lower limit on the proportion, of aromatic solvents. The rest of the blend consists of aliphatic solvents that do not pose such concerns for human health. A blend of aliphatic and aromatic hydrocarbons has useful solvent properties, especially for applying paints and older formulations of (coloured) printing inks, which cannot easily be matched by using only a combination of less harmful aliphatic solvents. The exact components in the aromatic portion of such commercial mixtures have probably changed over the years, with newer product codes having fewer or no suspected carcinogens in the mix. This is a good reason not to retain and use very old commercial solvents in a conservation studio or laboratory: the newer ones are tightly controlled as to the ingredients in the mixture, but a bottle that is decades old might include some carcinogenic solvents. (Another reason not to use mixtures with an old date marked on them is that some components might have evaporated off when the container was opened frequently, leaving a solvent mixture with slightly different properties from what is suggested by the name on the label.)

Phenyl alcohol or hydroxy benzene is known as phenol, with the formula shown in Figure 5.15. This is a volatile degradation product of some polymers, which can be used to identify them by sampling and then analysing their airspace rather than the plastic-based object itself.

Phenyl ethylene or vinyl benzene is commonly called styrene as shown in Figure 5.16, which is the basic unit of the polymer polystyrene. Styrenated resins are used as transparent casting resins for the making of cross-sections from paint samples, metals and many other objects that have coloured layers and coatings that have to be discovered and understood before a safe treatment can be selected to remove the top layer without affecting those underneath, even damaged or cracked layers that might channel in the solvent.

Another related compound that is sometimes very useful for one step of a complex treatment of an object, and also during the safe lifting of newly excavated archaeological material to take it to a conservation workspace for post-excavation stabilisation, is cyclododecane. Its formula is $(CH_2)_{12}$. While this appears at first glance like an aliphatic compound, a moment's thought indicates that such a molecule would be like a tail with no head, and hence that it must have some kind of ring structure: in fact, it looks like three six-carbon rings that fit together like tiles on a floor, as shown in Figure 5.17. This unique and close-packed arrangement gives the property that makes it useful: at cool room temperatures (and even more

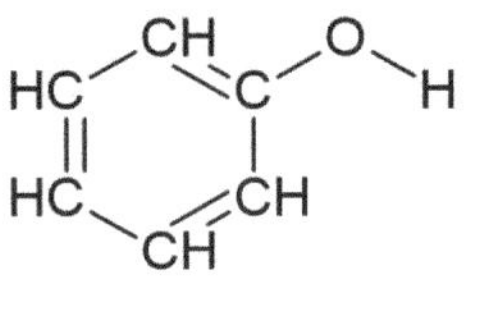

Figure 5.15 Structure of phenol.

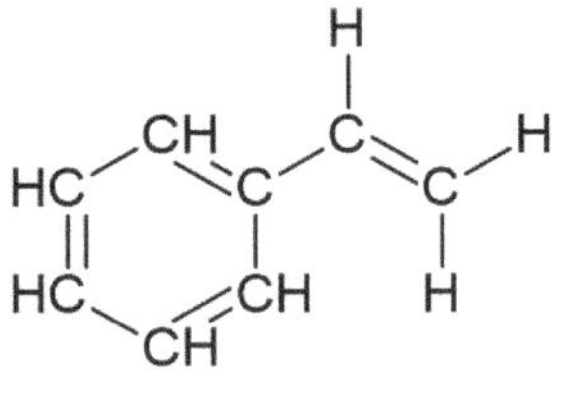

Figure 5.16 Structure of styrene.

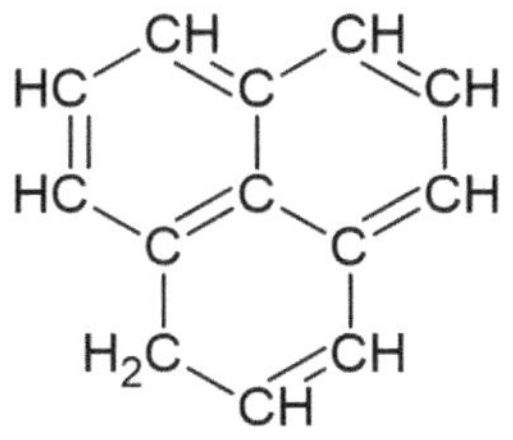

Figure 5.17 The structure of cyclododecane.

so in chilly weather during an excavation in winter) it is solid, but slight heating makes it liquid, and then it can be poured round a fragile object to give it gentle support, and to act as a barrier to the casting material or the supporting mould that is being constructed for the object and is almost-but-not-quite in contact with it. The cyclododecane soon evaporates and leaves the surface as a gas, and thus it is not likely to interact with the material being lifted, or treated later with different materials. It does however expose the conservators or archaeologists to a compound that can now be breathed in as a gas, and there have long been concerns in the field of conservation over the health and safety aspects of using this material. It is known to survive persistently in the environment.

All the solvents discussed above are petrochemicals. In recent years, 'green solvents' have been proposed for use in conservation, that are less harmful to human health than many of the aromatic solvents, or which are already made from plant-based sources, though for a non-conservation use. Another aspect of 'greenness' in solvents may be that they are not known to be harmful to the environment, which many aromatic compounds, especially those with high molecular weight, are. A material has even 'better' green credentials if it is **biodegradable**, meaning that it will break down to simple and harmless compounds in the outdoor environment when affected by light, water, or typical organisms found in the soil, or all of these. There is not (in 2023) a tight definition of which or how many such standards a solvent has to meet to justify the name **green solvent** within the field of conservation. A solvent might be green in one sense, and therefore be promoted as 'better' to use than the solvents in wide use by conservators today, but not green in all of these aspects. How to make use of greener materials or at least those known to be less harmful to the environment is a very active topic of discussion amongst conservators in many countries: solutions include substituting less toxic ones, cutting down waste, using products produced in the country of use, and re-using materials that were once thrown away after the first use – all while ensuring in the case of solvents that these less familiar materials have no bad long-term effects on objects. The newest information can be found in seminars and discussion forums hosted by many professional bodies and conservation programmes worldwide.

Research publications on solvent properties are now comparing the most widely suggested green solvent alternatives with the solvents typically used today to remove dirt from surfaces, or to swell coatings or varnishes to make their removal possible by gentle mechanical action. It is growing much easier to see which ones might be worth trying out in one's own workspace, for common conservation processes. One of the most frequently proposed green solvents is **limonene** (dipentene, $C_{10}H_{16}$), often the isomer called D-limonene which has been declared safe for consumption in food. This does have a six-carbon ring as Figure 5.18 shows, but it also has three alkyl groups attached, each with one carbon atom, making these carbon 'tails' as short as they could possibly be. It is the mixture of the ring and the short tails that makes its solvent properties interesting in conservation, helped by the fact that it is produced in large quantities by industry from citrus fruits rather than hydrocarbons, although for other purposes, including providing the scent in domestic cleaning products, and for cosmetics, and that it is biodegradable. Another green solvent which is often suggested is **ethyl lactate** (ethyl 2-hydroxypropanoate, $C_5H_{10}O_3$), which Figure 5.19 shows is not an aromatic solvent, but not a simple aliphatic solvent either. It can be produced from biomass, in other words from something that today is often simply disposed of, and it is biodegradable. Other green solvents have been proposed, but it is too soon for anyone to propose a direct substitute for propan-2-ol (isopropanol), or acetone, or white spirit, or any of the Shellsols® in use today.

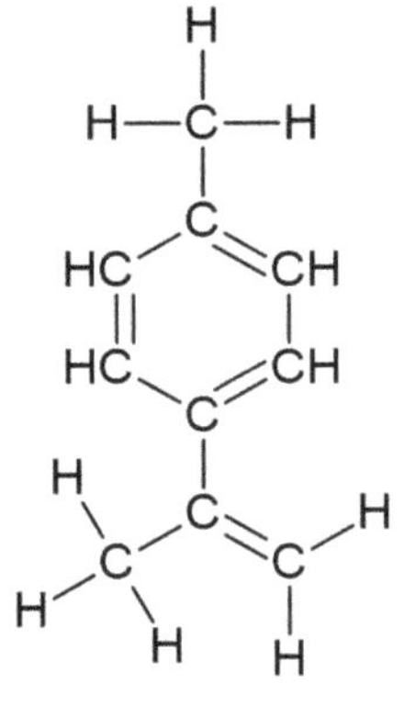

Figure 5.18 Structure of D-limonene.

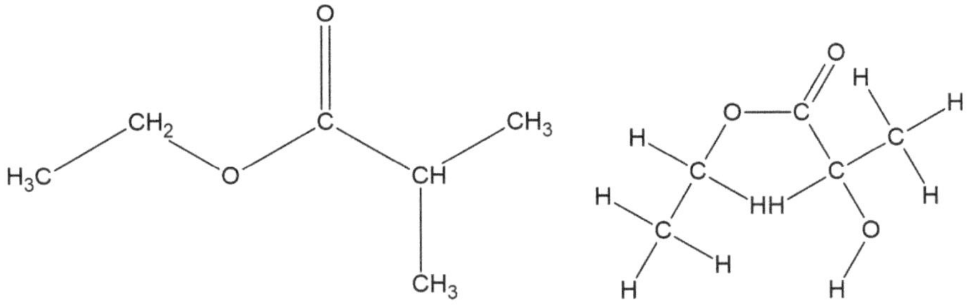

(a)

(b)

Figure 5.19 Abbreviated structure (a) and full structure (b) for ethyl lactate.

D   Conclusion

You should by now have a grasp of the theory of basic chemistry. In the book in the series which deals with the scientific aspects of *Cleaning*, you will start to use the theory in practice. The dirt on an object is frequently chemically very complex, and it is held on to the object by secondary bonds, which are related to the primary bonding mechanisms discussed in this book. The theory you have learnt should enable you to make an informed choice between the many available methods to break the bonds without damaging the object you are working on.

Answers to exercises

Chapter 3

1. (a) One atom of carbon and two atoms of oxygen
   (b) Two atoms of hydrogen, one atom of sulfur and four atoms of oxygen
   (c) Three atoms of carbon, six atoms of hydrogen and one atom of oxygen

2. (a) H_2O
   (b) NH_3
3. (a) $C_3H_8 + 5O_2 \rightarrow 3CO_2 + 4H_2O$
   (b) $2C_4H_{10} + 13O_2 \rightarrow 8CO_2 + 10H_2O$
4. $CaCO_3 + heat \rightarrow CaO + CO_2$
   Calcium carbonate + heat $\rightarrow$ calcium oxide + carbon dioxide
5. $CaCO_3 + H_2SO_4 \rightarrow CaSO_4 + H_2O + CO_2$
   calcium carbonate + sulfuric acid $\rightarrow$ calcium sulfate + water + carbon dioxide
6. Carbon dioxide
7. (a) CH_4 methane with molecular weight $1 \times 12 + 4 \times 1 = 16$
   (b) O_2 oxygen with molecular weight $1 \times 16 = 32$
   (c) H_2O water with molecular weight $2 \times 1 + 16 = 18$

Chapter 4

1. Check answer against Table 4.1
2. To enable each atom to make its correct number of links, the answers must be:
   (a) H_2 is H–H
   (b) N_2 is N–H

   (c) ammonia is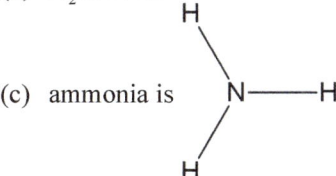

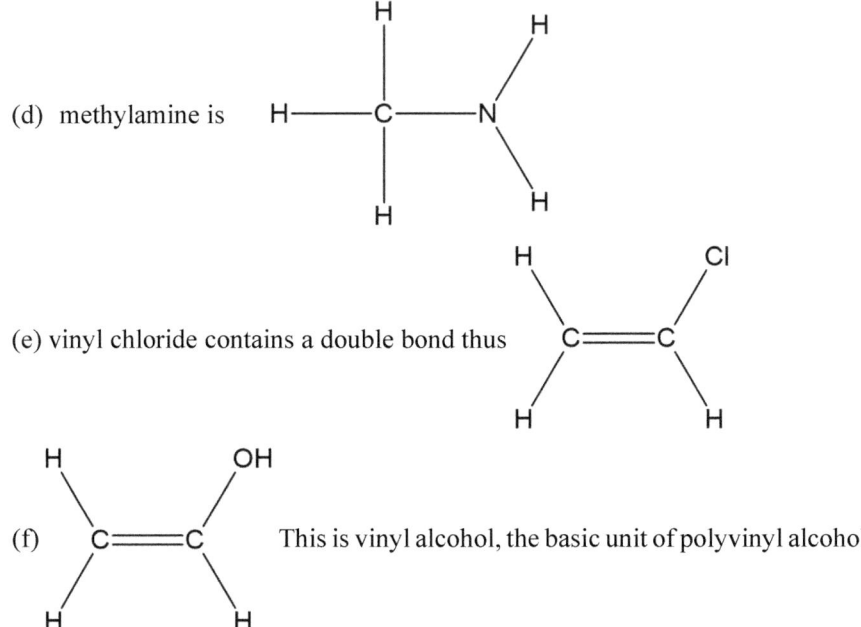

(d) methylamine is

(e) vinyl chloride contains a double bond thus

(f)   This is vinyl alcohol, the basic unit of polyvinyl alcohol

3. Nitrogen atoms have five electrons outside the helium (two-electron) shell. To become like neon each atom must acquire a share in three more electrons. So the structures are as follows:

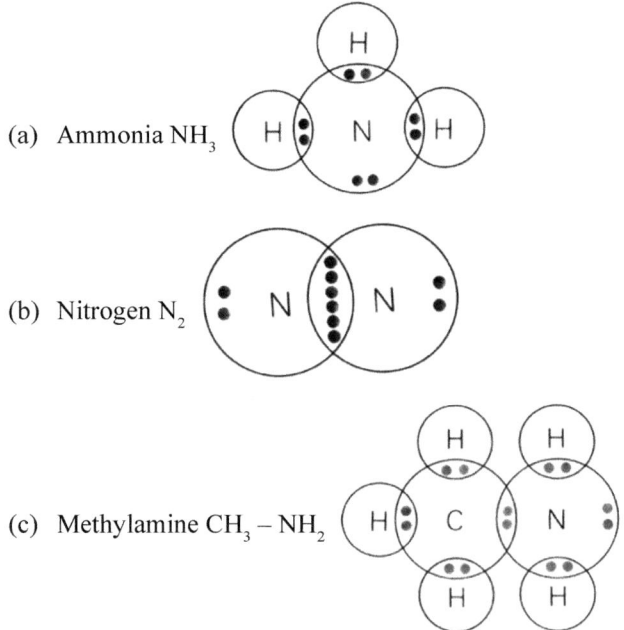

(a) Ammonia NH_3

(b) Nitrogen N_2

(c) Methylamine $CH_3 - NH_2$

Chapter 5

1. (a) Methyl alcohol, or methanol
   (b) Propyl alcohol, or propanol
   (c) Ethylamine
   (d) dimethyl ether

Periodic table of the elements

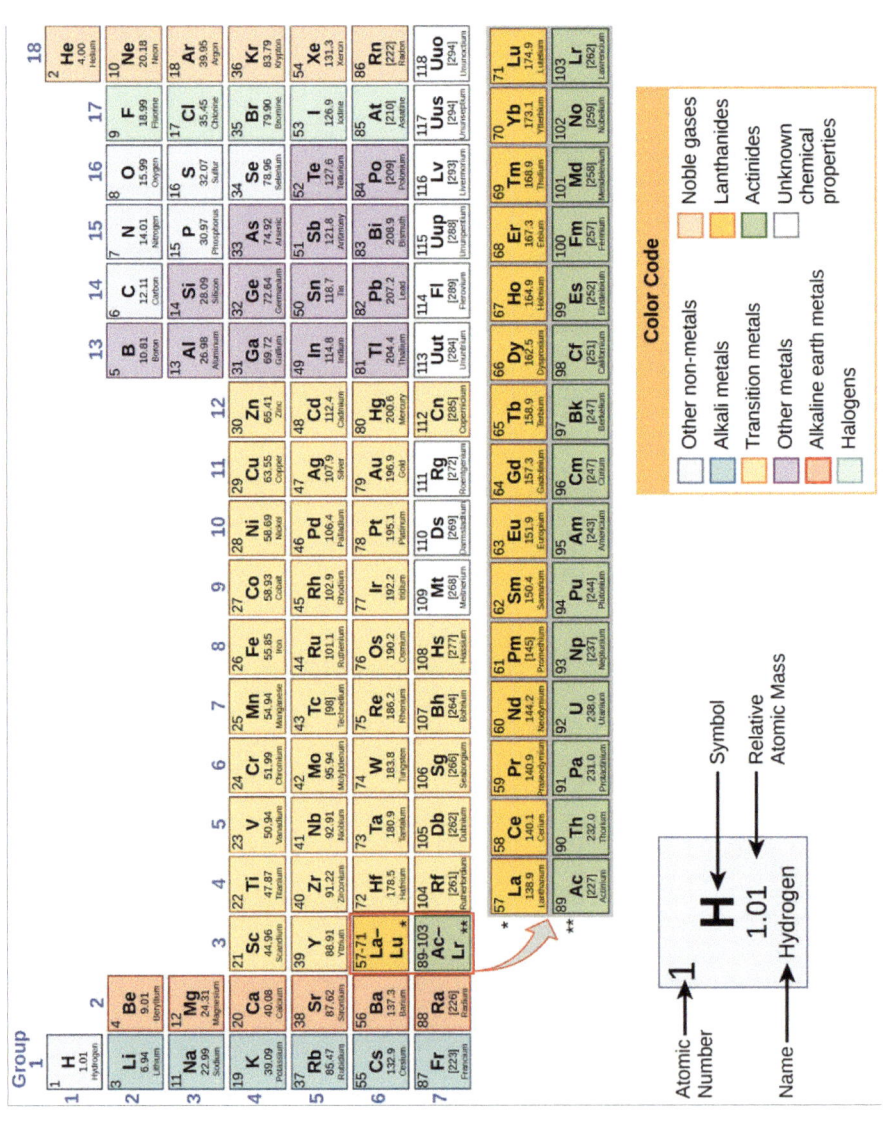

Further reading

AIC wiki on instrumental analysis. n.d. https://www.conservation-wiki.com/wiki/Category:Instrumental_Analysis)

Artioli, G. (ed.) 2010. *Scientific Methods and Cultural Heritage: An Introduction to the Application of Materials Science to Archaeometry and Conservation Science.* Oxford: Oxford University Press.

Broecke. L. (tr.) 2015. Cennino D'Andrea Cennini's *Il Libro dell'Arte /The Craftsman's Handbook.* London: Archetype Publications.

Fife, G. L. (ed.) 2021. *Greener Solvents in Conservation.* London: Archetype Publications. https://www.siconserve.org/greener-solvents/greener-solvents-hand-book/.

Henderson, J. 2000. *The Science and Archaeology of Materials.* Abingdon: Routledge.

Hoenig, S. 2020. *Basic Chemical Concepts and Tables.* Boca Raton, FL: CRC Press (Taylor & Francis Group).

Horie, C. V. 2010. *Materials for Conservation*, 2nd edn. Abingdon: Routledge.

Kingery, W. (ed.) 1996. *Learning from Things: Method and Theory of Material Culture Studies.* Washington, D.C.: Smithsonian Institution Press.

Matteini, M., R. Mazzeo and A. Moles 2016. *Chemistry for Restoration: Painting and Restoration Materials.* Florence: Nardini Editore.

May, E. and M. Jones (eds) 2006. *Conservation Science: Heritage Materials.* Cambridge: RSC Publishing.

Mills, J. and R. White 2012. *The Organic Chemistry of Museum Objects.* Abingdon: Routledge.

Pollard, A. M. and C. Heron 2008. *Archaeological Chemistry*, 2nd edn. Cambridge: Royal Society of Chemistry.

Price, T. D. and J. H. Burton (eds) 2011. *An Introduction to Archaeological Chemistry.* New York and London: Springer.

Royal Society of Chemistry 2021. Interactive Periodic Table. Available from: http://www.rsc.org/periodic-table.

STiCH, Tools for Informed Sustainable Choices (https://stich.culturalheritage.org).

Stuart, B. 2007. *Analytical Techniques in Materials Conservation.* London: Wiley.

Studies in Conservation (online and print). Free access to IIC members through the IIC website www.iiconservation.org/publications. Pay-per-view access at https://www.tandfonline.com/loi/ysic20. 2015 and earlier volumes available without payment: https://www.jstor.org/journal/studcons.

Wei, W. 2021. *Art Conservation: Mechanical Properties and Testing of Materials*, Jenny Stanford Publishing. [2021 hardcover and ebook]

Index